Tucholsky   Wagner   Zola   Scott   Sydow   Freud   Schlegel
Turgenev   Wallace   Fonatne
Twain   Walther von der Vogelweide   Fouqué   Friedrich II. von Preußen
    Weber   Freiligrath   Frey
    Weiße Rose   Kant   Ernst   Frommel
Fechner   Fichte   von Fallersleben   Richthofen
    Engels   Fielding   Hölderlin   Tacitus   Dumas
Fehrs   Faber   Eichendorff
    Maximilian I. von Habsburg   Flaubert   Eliasberg   Ebner Eschenbach
Feuerbach   Ewald   Fock   Eliot   Zweig   Vergil
    Goethe   Elisabeth von Österreich   London
Mendelssohn   Balzac   Shakespeare   Dostojewski   Ganghofer
    Trackl   Lichtenberg   Rathenau   Hambruch   Doyle   Gjellerup
Mommsen   Stevenson   Tolstoi   Lenz   Hanrieder   Droste-Hülshoff
    Thoma   von Arnim   Hägele   Hauff   Humboldt
Dach   Verne   Rousseau   Hagen   Hauptmann   Gautier
    Reuter   Defoe   Hebbel   Baudelaire
Karrillon   Garschin   Descartes   Hegel   Kussmaul   Herder
    Damaschke   Dickens   Schopenhauer   Rilke   George
Wolfram von Eschenbach   Darwin   Melville   Grimm   Jerome   Bebel   Proust
Bronner   Campe   Horváth   Aristoteles   Barlach   Voltaire   Federer   Herodot
Bismarck   Vigny   Gengenbach   Heine
    Storm   Casanova   Tersteegen   Gilm   Grillparzer   Georgy
    Chamberlain   Lessing   Langbein   Gryphius
Brentano   Claudius   Schiller   Lafontaine   Kralik   Iffland   Sokrates
Strachwitz   Bellamy   Schilling
    Katharina II. von Rußland   Gerstäcker   Raabe   Gibbon   Tschechow
Löns   Hesse   Hoffmann   Gogol   Wilde   Gleim   Vulpius
    Luther   Heym   Hofmannsthal   Klee   Hölty   Morgenstern   Goedicke
    Roth   Heyse   Klopstock   Puschkin   Homer   Kleist
Luxemburg   La Roche   Horaz   Mörike   Musil
    Machiavelli   Kierkegaard   Kraft   Kraus
Navarra   Aurel   Musset   Lamprecht   Kind   Kirchhoff   Hugo   Moltke
Nestroy   Marie de France   Laotse   Ipsen   Liebknecht
    Nietzsche   Nansen   Ringelnatz
    Marx   Lassalle   Gorki   Klett   Leibniz
von Ossietzky   May   vom Stein   Lawrence   Irving
    Petalozzi   Knigge
    Platon   Pückler   Michelangelo   Kock   Kafka
    Sachs   Poe   Liebermann
    de Sade   Praetorius   Mistral   Zetkin   Korolenko

The publishing house tredition has created the series **TREDITION CLASSICS**. It contains classical literature works from over two thousand years. Most of these titles have been out of print and off the bookstore shelves for decades.

The book series is intended to preserve the cultural legacy and to promote the timeless works of classical literature. As a reader of a **TREDITION CLASSICS** book, the reader supports the mission to save many of the amazing works of world literature from oblivion.

The symbol of **TREDITION CLASSICS** is Johannes Gutenberg (1400 – 1468), the inventor of movable type printing.

With the series, tredition intends to make thousands of international literature classics available in printed format again – worldwide.

All books are available at book retailers worldwide in paperback and in hardcover. For more information please visit: www.tredition.com

tredition was established in 2006 by Sandra Latusseck and Soenke Schulz. Based in Hamburg, Germany, tredition offers publishing solutions to authors and publishing houses, combined with worldwide distribution of printed and digital book content. tredition is uniquely positioned to enable authors and publishing houses to create books on their own terms and without conventional manufacturing risks.

For more information please visit: www.tredition.com

Handbook of Embroidery

L. Higgin

Imprint

This book is part of the TREDITION CLASSICS series.

Author: L. Higgin
Cover design: toepferschumann, Berlin (Germany)

Publisher: tredition GmbH, Hamburg (Germany)
ISBN: 978-3-8491-4922-2

www.tredition.com
www.tredition.de

Copyright:
The content of this book is sourced from the public domain.

The intention of the TREDITION CLASSICS series is to make world literature in the public domain available in printed format. Literary enthusiasts and organizations worldwide have scanned and digitally edited the original texts. tredition has subsequently formatted and redesigned the content into a modern reading layout. Therefore, we cannot guarantee the exact reproduction of the original format of a particular historic edition. Please also note that no modifications have been made to the spelling, therefore it may differ from the orthography used today.

PREFACE.

In drawing up this little "Handbook of Embroidery" we do not pretend to give such complete technical directions as would enable a beginner in this beautiful art to teach herself; because learning without practical lessons must be incomplete, and can only lead to disappointment.

We have sought, therefore, only to respond to the inquiries we are constantly receiving, and to supply useful hints to those who are unable to avail themselves of lessons, and are forced to puzzle over their difficulties without help from a trained and experienced embroiderer; at the same time, the rules we have laid down and the directions we have given may serve to remind those who have passed through the classes, of many little details which might easily be forgotten when the lessons are over, though so much of the success of embroidery depends upon them.

[Pg vi] We have given a short description of the most useful stitches, and have pointed out their applicability to different styles of work; we have named the various materials which are best suited as grounds for embroidery, and the silks, filoselles, crewels, &c., which are most commonly employed, with practical rules for their use in the best and most economical manner.

Also we have given such plain directions as to stretching, framing, and cleaning the work as are possible in a limited space, and without practical illustration. We venture to hope we have thus supplied a want that has been long felt by those who interest themselves in the art in which Englishwomen once excelled, but which had languished of late years, and almost died out amongst us, though it has always been taught in many continental cities, where embroideries have never ceased to be required for church decoration.

We have abstained from giving any directions as to the tracing of designs upon material, for two sufficient reasons: firstly, that the Royal School of Art-Needlework has never supplied designs alone, or in any other form than as prepared work; and secondly, that having made experiments with all the systems that have been brought out for "stamping," ironing from transfer-papers, or with tracing powder, it has been found that designs can only be artistically and well traced on material by hand painting. Those ladies who can design and paint their own patterns for embroidery are independent of assistance, and to those who are unable to do so we cannot recommend any of the methods now advertised.

[Pg vii] It has been thought unnecessary to enter into the subject of ecclesiastical embroidery at present. This has been so thoroughly revived in England, and practised in such perfection by sisterhoods—both Anglican and Roman Catholic—as well as by some of the leading firms of church decorators, that we have not felt ourselves called upon to do more than include it in our course of lessons.

The æsthetic side of our subject we have purposely avoided, as it would lead us further than this purely technical guide-book pretends to go. But we propose shortly to bring out a second part devoted to design, composition, colour, and the common-sense mode of treating decorative Art, as applied to wall-hanging, furniture, dress, and the smaller objects of luxury.

We shall examine and try to define the principles which have guided Eastern and Western embroideries at their best periods, hoping thus to save the designers of the future from repeating exploded experiments against received canons of good taste; checking, if we can, the exuberance of ignorant or eccentric genius, but leaving room for originality.

Mrs. Dolby, who by her presence and her teaching helped Lady Welby to start the Royal School of Art-Needlework, has left behind her a most valuable guide for mediæval work in her "Church Embroidery, Ancient and Modern," which will always be a first-class authority.

The Author and the Editor of this handbook are equally impressed with the responsibility they have [Pg viii] undertaken in formulating rules for future embroiderers. They have consulted all acknowledged authorities, and from them have selected those which the teachers in the Royal School of Art-Needlework have found the most practical and instructive.

Should any of their readers favour them with hints or criticisms, or give them information as to pieces of embroidery worth studying, or stitches not here named, any such communications will be gratefully received and made use of in future editions.

The Editor.

TABLE OF CONTENTS.

CHAPTER I.
Of Implements and Materials used in Modern Embroidery.

Needles
Scissors
Prickers, &c.
Crewels
Tapestry Wool
Arrasene
Embroidery or Bobbin Silk
Rope Silk
Fine Silk
Purse Silk
Raw or Spun Silk
Vegetable Silk
Filoselle
Tussore
Gold
Japanese Gold Thread
Chinese Gold
Gold and Silver Passing
[Pg x] Bullion or Purl
Spangles
Plate
Recipes for Preserving Gold

CHAPTER II.

Textile Fabrics used as Grounds for Embroidery.

Linens
Flax
Twill
Kirriemuir Twill
Sailcloth
Oatcake Linen
Oatmeal Linen
Smock Linen
Bolton, or Workhouse Sheeting
Satins and Silks
Silk Sheeting
Tussore and Corah Silks
Plain Tapestries
Brocatine
Cotton and Woollen
Velveteen
Utrecht Velvet
Velvet Cloth
Felt
Diagonal Cloth
Serge
Soft, or Super Serge
Cricketing Flannel
Genoa or Lyons Velvet
Silk Velvet Plush
Cloths of Gold and Silver

CHAPTER III.
Stitches.
Stem Stitch
Split Stitch
Satin Stitch
Blanket Stitch
Button-hole Stitch
Knotted Stitch
Chain Stitch
Twisted Chain
Feather Stitch

CHAPTER IV.
Frames and Framing

CHAPTER V.
Stitches used in Frame Embroidery.
Feather Stitch
Couching or Laid Embroidery
Net-patterned Couching
Brick Stitch
Diaper Couchings
Basket Stitch
Spanish Embroidery
Cross Stitch
Simple Cross Stitch
[Pg xii] Persian Cross Stitch
Burden Stitch
Stem Stitch
Japanese Stitch

Tambour Work
Opus Anglicum
Cut Work
Inlaid Appliqué
Onlaid Appliqué
Gold Embroidery
Backing
Stretching and Finishing
Embroidery Paste
Cleaning

ILLUSTRATIONS.

Description of the Plates

Sixteen Plates, containing 24 Designs

[Pg 1]

HANDBOOK OF EMBROIDERY.

CHAPTER I.

OF MATERIALS AND IMPLEMENTS USED IN MODERN EMBROIDERY.

IMPLEMENTS.

Needles.—The best "embroidery needles" for ordinary crewel handwork are Nos. 5 and 6. For coarse "sailcloth," "flax," or "oatcake," No. 4. For frame embroidery, or very fine handwork, the higher numbers, from 7 to 10.

It is a mistake to use too fine a needle. The thread of crewel or silk should always be able to pass loosely into the eye, so as not to require any pulling to carry it through the material.

Scissors should be finely pointed, and very sharp.

[Pg 2] **Thimbles** which have been well worn, and are therefore smooth, are best. Some workers prefer ivory or vulcanite. Two thimbles should be used for framework.

Prickers are necessary for piercing holes in gold embroidery, and also for arranging the lie of the thread in some forms of couching.

[Pg 3]

MATERIALS.

CREWELS, AND HOW TO USE THEM.

Crewel should be cut into short threads, never more than half the length of the skein. If a long needleful is used, it is not only apt to pull the work, but is very wasteful, as the end of it is liable to become frayed or knotted before it is nearly worked up. If it is necessary to use it double (and for coarse work, such as screen panels on sailcloth, or for embroidering on Utrecht velvet, it is generally better doubled), care should be taken never to pass it through the eye of the needle, knotting the two ends; but two separate threads of the length required should be passed together through the needle.

Crewel should not be manufactured with a twist, as it makes the embroidery appear hard and rigid; and the shades of colour do not blend into each other so harmoniously as when they are untwisted.

In crewels of the best quality the colours are perfectly fast, and will bear being repeatedly washed, provided no soda or washing-powder is used. Directions for cleaning [Pg 4] crewel work are given later; but it should not be sent to an ordinary laundress, who will most certainly ruin the colours.

Crewel is suitable for embroidery on all kinds of linen—on plain or diagonal cloth, serge, flannel, &c. It is also very effective when used in conjunction with embroidery silk, or filoselle, either in conventional designs, or where flowers are introduced. The leaves may be worked in crewels, and the flowers in silk, or the effect of the crewels increased by merely touching up the high lights with silk.

Tapestry Wool is more than twice the thickness of crewel, and is used for screen panels, or large curtain borders, where the work is coarse, and a good deal of ground has to be covered. It is also used for bath blankets and carriage and sofa rugs. Tapestry wool is not yet made in all shades.

Fine crewels are used for delicately working small figures, d'oyleys, &c.; but there is also a difficulty about obtaining these in all shades, as there is not much demand for them at present.

Arrasene is a new material. It is a species of worsted chenille, but is not twisted round fine wire or silk, like ordinary chenille; though it is woven first into a fabric, and then cut in the same manner. It serves to produce broad effects for screen panels, or borders, and has a very soft, rich appearance when carefully used. It is made also in silk; but this is inferior to worsted arrasene, or the old-fashioned chenille.

[Pg 5]

SILKS.

"**Embroidery**," or **Bobbin Silk**, which has now almost superseded floss, is used for working on satin and silk, or for any fine work. It is made in strands, each of which has a slight twist in it to prevent its fraying as floss does. As this silk is required in all varieties of thickness, it is manufactured in what is technically called "rope," that is, with about twelve strands in each thread. When not "rope" silk, it is in single strands, and is then called "fine" silk. As it is almost always necessary to use several strands, and these in varying number, according to the embroidery in hand, the rope silk has to be divided, or the fine doubled or trebled, as the case may be.

If rope silk is being used, the length required for a needleful must be cut and passed carefully between finger and thumb once or twice, that it may not be twisted. It should then be carefully separated into the number of strands most suitable for the embroidery in hand; for ordinary work three is about the best number.

These must be threaded together through the needle, care being taken not to tangle the piece of "rope" from which they have been detached. There need be no waste [Pg 6] if this operation is carefully done, as good silk will always divide into strands without fraying.

In using "fine silk," one length must be cut first, then other strands laid on it,—as many as are needed to form the thickness required. They should be carefully laid in the same direction as they leave the reel or card. If placed carelessly backwards and forwards, they are sure to fray, and will not work evenly together. With silk still more than with crewel, it is necessary to thread all the strands through the needle together, never to double one back, and never to make a knot.

It is intended in future to do away with this distinction between "rope" and "fine" silk, and to have it all manufactured of one uniform thickness, which will consist of eight strands of the same quality as the "fine" silk at present in use. As it will, however, still be necessary to divide the thread, and even perhaps occasionally to double it, the directions given above will be useful.

Purse Silk is used sometimes for diapering, and in rare cases in ordinary embroidery, where a raised effect is required.

Raw or **spun silk** is a soft untwisted cream-coloured silk, used for daisies and other simple white flowers, or in outlining. It is much cheaper than embroidery silk or filoselle.

Vegetable Silk (so-called) is not used or sold by the Royal School.

[Pg 7] **Filoselle**, when of good quality, is not, as some people suppose, a mixture of silk and cotton. It is pure silk, but of an inferior quality; and therefore cheaper. It answers many of the purposes of bobbin silk, but is not suitable for fine embroidery on silk or satin fabrics. It should be used also in strands, and the same remarks hold good with regard to its not being doubled, but cut in equal lengths.

Tussore.—Interesting experiments have recently been made with the "Tussore," or "wild silk" of India, which bids fair to create a revolution in embroidery. Not only can it be produced for less than half the price of the "cultivated silk" of Italy, China, or Japan, but it also takes the most delicate dyes with a softness that gives a peculiarly charming effect. It can scarcely be said to be in the market as yet, but in all probability before this work is through the press it will have become an important element in decorative needlework. It is much less glossy than cultivated silk.

GOLD THREAD, &c.

"**Japanese gold thread**," which has the advantage of never tarnishing, is now extremely difficult to obtain. Being made of gilt paper twisted round cotton thread, it cannot be drawn through the material by the needle; but must in all cases be laid on, and stitched down with a fine yellow silk, known as "Maltese," or "Horse-tail."

"**Chinese gold**" is manufactured in the same manner as the Japanese; but being of a much redder colour is not so satisfactory in embroidery unless a warm shade is desirable for a particular work.

Gold and silver passing, a very fine kind of thread, can either be used for working through the material, or can be laid on like the Japanese gold. They are suitable for "raised gold or silver embroidery."

Bullion, or Purl, is gold or silver wire made in a series of continuous rings, like a corkscrew. It is used in ecclesiastical work, for embroidering official and military uniforms, and for heraldic designs. It should be cut into the required lengths—threaded on the needle [Pg 9] and fastened down as in bead-work. Purl is sometimes manufactured with a coloured silk twisted round the metal though not concealing it, and giving rich tints to the work.

Spangles were anciently much used in embroidery, and were sometimes of pure gold. They are but little used now.

Plate consists of narrow plates of gold or silver stitched on to the embroidery by threads of silk, which pass over them.

The French and English gold thread is made of thin plates of metal cut into strips, and wound round strands of cotton in the same manner as the Japanese gold. If the metal is real, the cost is of course great. It is sold by weight, gold being about 20s. per oz., and silver, 10s. per oz. In addition to its superiority in wear, it has this advantage, that old gold or silver thread is always of intrinsic value, and may be sold at the current price of the metal whatever state it may be in. Many varieties of gilt thread are manufactured in France and England, which may be used when the great expense of "real gold" is objected to. But although it looks equally well at first, it soon becomes tarnished, and spoils the effect of the embroidery. Gold and silver threads are difficult to work with in England, and especially in London, as damp and coal-smoke tarnish them almost before the work is out of the frame. Mrs. Dolby recommends cloves being placed in the papers in which they are kept.

[Pg 10]

RECIPES FOR PRESERVING GOLD.

We give here two recipes, which may be found serviceable. They are from different sources; the first is a very old one. They may preserve gold for a certain time.

1. Isinglass dissolved in spirits of wine and brushed over the thread or braid, which should be hung over something to dry, and not touched with the hand.

2. Spirits of wine and mastic varnish mixed very thin and put on in the same way with a brush.

[Pg 11]

CHAPTER II.

TEXTILE FABRICS USED AS GROUNDS FOR EMBROIDERY.

LINENS.

There are many varieties of unglazed, half-bleached linens, from that thirty-six and forty inches wide, used for chair-back covers, to that ninety inches wide, used for large table-covers, curtains, &c. There are also endless varieties of fancy linens, both of hand and power-loom weaving, for summer dresses, for bed furniture, chair-back covers, table-cloths, &c.

Flax is the unbleached brown linen, often used for chair-back covers.

Twill is a thick linen suitable for coverings for furniture.

Kirriemuir Twill is a fine twilled linen made at Kirriemuir, and is good for tennis aprons, dresses, curtains, &c.

[Pg 12] **Sailcloth** is a stout linen, of yellow colour, and is only suitable for screen panels.

Oatcake Linen, so called from its resemblance to Scotch oatcake, has been popular for screen panels or washstand backs. It is very coarse and rough.

Oatmeal Linen is finer and of a greyer tone. It is also used for screens, and for smaller articles.

Smock Linen is a strong even green cloth. It makes an excellent ground for working screens, and is also used for tennis aprons.

Crash.—Properly speaking, the name "*crash*" is only applied to the coarse Russian home-spun linen, which has been such a favourite from the beauty of its tone of colour. It is, however, erroneously applied to all linens used for embroidery, whether woven by hand-loom or machinery; and this confusion of names frequently leads to mistakes. Crash is almost always very coarse, is never more than eighteen inches wide, and cannot be mistaken for a machine-made fabric. It is woven by the Russian peasants in their own homes, in lengths varying from five to ten yards, and, therefore, though sent over in large bales, it is very difficult to find two pieces among a hundred that in any way match each other.

Bolton, or Workhouse Sheeting, is a coarse twilled cotton fabric, seventy-two inches wide, of a beautiful soft creamy colour, which improves much in washing. It is [Pg 13] inexpensive, and an excellent ground for embroidery, either for curtains, counterpanes, chair coverings, or for ladies' dresses, or tennis aprons.

It resembles the twilled cotton on which so much of the old crewel embroidery was worked in the seventeenth century, and is one of the most satisfactory materials when of really good quality.

All descriptions of linen, except the "oatcake" and "sailcloth," can be embroidered in the hand.

[Pg 14]

TEXTILE FABRICS.

SATINS AND SILKS.

Satins and Silks can only be embroidered in a frame. Furniture satins of stout make, with cotton backs, may be used without backing; but ordinary dress satins require to have a thin cotton or linen backing to bear the strains of the work and framing. Nothing is more beautiful than a rich white satin for a dress embroidered in coloured silks.

For fans, a very fine, closely woven satin is necessary, as it will not fold evenly unless the satin is thin; and yet it must be rich enough to sustain the fine embroidery, without pulling, or looking poor. A special kind of satin is made for the manufacture of fans, and none other is available.

"**Silk Sheeting**" of good quality, "*Satin de Chine*" and other silk-faced materials of the same class, may either be embroidered in the hand, or framed; but for large pieces of work a frame is essential. These materials are suitable for curtains, counterpanes, piano coverings, [Pg 15] or panels, and indeed for almost any purpose. The finer qualities are very beautiful for dresses, as they take rich and graceful folds, and carry embroidery well.

Tussore and Corah Silks are charming for summer dresses, light chair-back covers, or embroidered window blinds. They will only bear light embroidering in silk or filoselle.

Within the last year successful experiments have been made in dyeing these Indian silks in England. The exact shades which we admire so much in the old Oriental embroideries have been reproduced, with the additional advantage of being perfectly fast in colour.

Nothing can be more charming as lining for table-covers, screens, curtains, &c.; and they are rather less expensive than other lining silks.

The fabrics known as **Plain Tapestries** are a mixture of silk and cotton, manufactured in imitation of the handworked backgrounds so frequent in ancient embroideries—especially Venetian. Almost all the varieties of *Opus Pulvinarium*, or cushion stitch, have been reproduced in these woven fabrics.

Brocatine is a silk-faced material, woven to imitate couched embroidery. The silk is thrown to the surface and is tied with cotton threads from the back.

As ground for embroidery it has an excellent effect.

[Pg 16]

TEXTILE FABRICS.

COTTONS AND WOOLLENS.

Velveteen, if of good quality, makes an excellent ground for screen panels, chair-covers, portières, curtains, borders, &c. It can be worked in the hand if the embroidery be not too heavy or large in style.

Utrecht Velvet is only suitable for coarse crewel or tapestry wool embroidery. It is fit for curtain dados or wide borderings.

Velvet Cloth is a rich plain cloth, finished without any gloss. It is a good ground for embroidery, either for curtains or altar-cloths. It is two yards wide.

Felt is sometimes used for the same purposes, but does not wear nearly so well, and is difficult to work.

Diagonal Cloth can be worked either in the hand or frame, although it is always much better in the latter. It is used for table-covers, curtains, chair-seats, &c.

[Pg 17] **Serge** is usually made thirty-six inches wide. It has long been in favour for curtains, small table-covers, dresses, &c. It can now be obtained at the school fifty-four inches wide, in many shades.

Soft or Super Serge, also fifty-four inches wide, is an excellent material, much superior in appearance to diagonal cloth, or to the ordinary rough serge. It takes embroidery well.

Cricketing flannel is used for coverlets for cots, children's dresses, and many other purposes. It is of a beautiful creamy colour, and is a good ground for fine crewel or silk embroidery. It need not be worked in a frame.

Genoa or Lyons Velvet makes a beautiful ground for embroidery; but it can only be worked in a frame, and requires to be "backed" with a thin cotton or linen lining, if it is to sustain any mass of embroidery. For small articles, such as sachets or casket-covers, when the work is fine and small, the backing is not necessary. Screen panels of velvet, worked wholly in crewels, or with crewel brightened with silk, are very effective. Three-piled velvet is the best for working upon, but is so expensive that it is seldom asked for.

Silk Velvet Plush (a new material) can only be used in frame work, and must be backed. It is useful in "appliqué" from the many beautiful tones of colour it takes. As a ground for silk or gold embroidery it is also very good.

[Pg 18]

TEXTILE FABRICS.

GOLD AND SILVER CLOTH.

Cloth of Gold or Silver is made of threads of silk woven with metal, which is thrown to the surface. In its best form it is extremely expensive, varying from £4 to £6 per yard, according to the weight of gold introduced. Cloth of silver is generally £3 the yard.

Inferior kinds of these cloths are made in which silk largely predominates, and shows plainly on the surface. They are frequently woven in patterns, such as diaper or diagonal lines, with a tie of red silk, in imitation of the diaper patterns of couched embroidery.

They are chiefly used in ecclesiastical or heraldic embroidery; their great expense preventing their general use.

[Pg 19]

CHAPTER III.

STITCHES USED IN HAND EMBROIDERY AS TAUGHT AT THE ROYAL SCHOOL OF ART-NEEDLEWORK.

To avoid pulling or puckering the work, care should be taken — firstly, that the needle is not too small, so as to require any force in drawing it through the material; secondly, the material must be held in a convex position over the fingers, so that the crewel or silk in the needle shall be looser than the ground; and thirdly, not to use too long needlefuls. These rules apply generally to all handworked embroideries.

STITCHES.

Stem Stitch.—The first stitch which is taught to a beginner is "stem stitch" (wrongly called also, "crewel stitch," as it has no claim to being used exclusively in crewel embroidery). It is most useful in work done in [Pg 20] the hand, and especially in outlines of flowers, unshaded leaves, and arabesque, and all conventional designs.

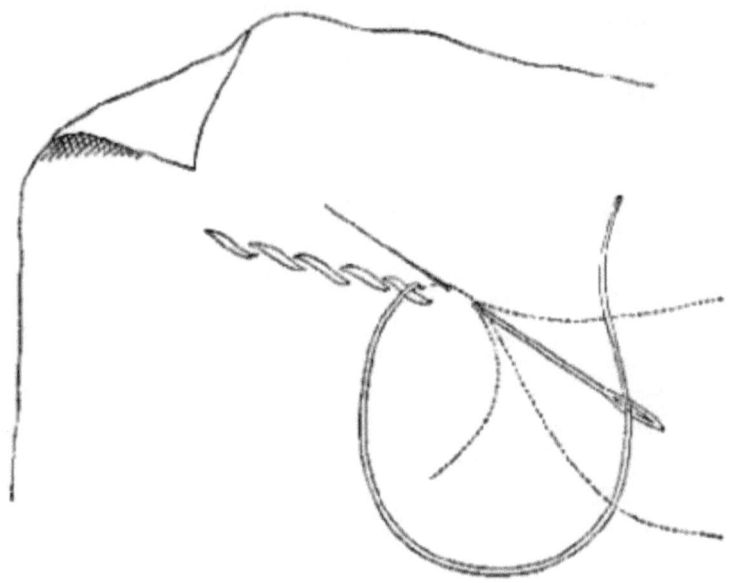

No. 1.—Stem Stitch.

It may be best described as a long stitch forward on the surface, and a shorter one backward on the under side of the fabric, the stitches following each other almost in line from left to right. The effect on the wrong side is exactly that of an irregular back-stitching used by dressmakers, as distinguished from regular stitching. A leaf worked in outline should be begun at the lower or stalk end, and worked round the right side to the top, taking care that the needle is to the left of the thread as it is drawn out. When the point of the leaf is reached, it is best to reverse the operation in working down the left side towards the stalk again, so as to keep the needle to the right of the thread instead of to the left, as in going up.

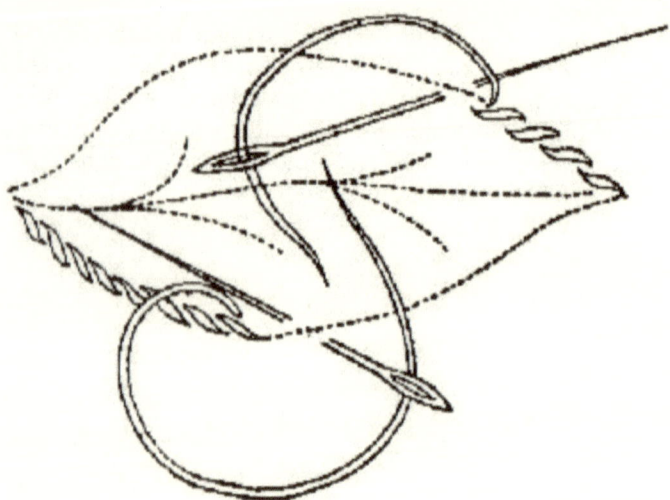

No. 2.

The reason of this will be easily understood: we will suppose the leaf to have a slightly serrated edge (and there is no leaf in nature with an absolutely smooth one). It will be found that in order to give this ragged appearance, it is necessary to have the points at which the insertions of the needle occur on the outside of the leaf: whereas if the stem stitch were continued down the left side, exactly in the same manner as in ascending the right, we should have the ugly anomaly of a leaf outlined thus:—

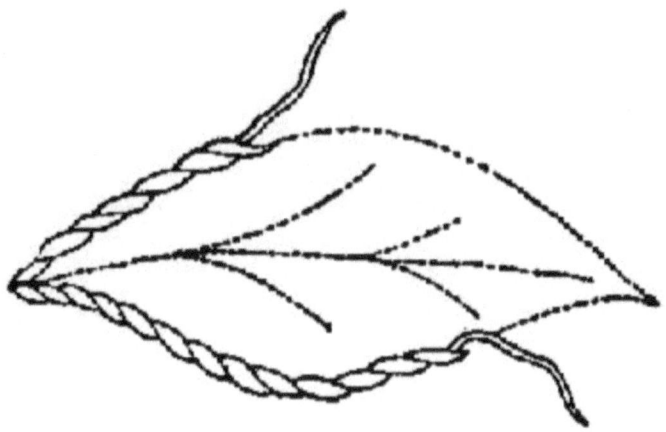

No. 3.

If the leaf is to be worked "solidly," another row of stem stitching must be taken up the centre of it (unless it be a very narrow leaf), to the top. The two halves of the leaf must then be filled in, separately, with close, even rows of stem stitch, worked in the ordinary way, [Pg 22] with the needle to the left of the thread. This will prevent the ugly ridge which remains in the centre, if it is worked round and round the inside of the outline. Stem stitch must be varied according to the work in hand. If a perfectly even line is required, care must be taken that the direction of the needle when inserted is in a straight line with the preceding stitch. If a slight serrature is required, each stitch must be sloped a little by inserting the needle at a slight angle, as shown in the illustration. The length of the surface stitches must vary to suit the style of each piece of embroidery.

Split Stitch is worked like ordinary "stem," except that the needle is always brought up *through* the crewel or silk, which it splits, in passing.

The effect is to produce a more even line than is possible with the most careful stem stitch. It is used for delicate outlines. Split stitch is rarely used in hand embroidery, being more suitable for frame work: but [Pg 23] has been described here as being a form of stem stitch. The effect is somewhat like a confused chain stitch.

Satin Stitch—*French Plumetis*—is one of those chiefly used in white embroidery, and consists in taking the needle each time back again almost to the spot from which it started, so that the same amount of crewel or silk remains on the back of the work as on the front. This produces a surface as smooth as satin: hence its name. It is chiefly used in working the petals of small flowers, such as "Forget-me-nots," and in arabesque designs where a raised effect is wanted in small masses.

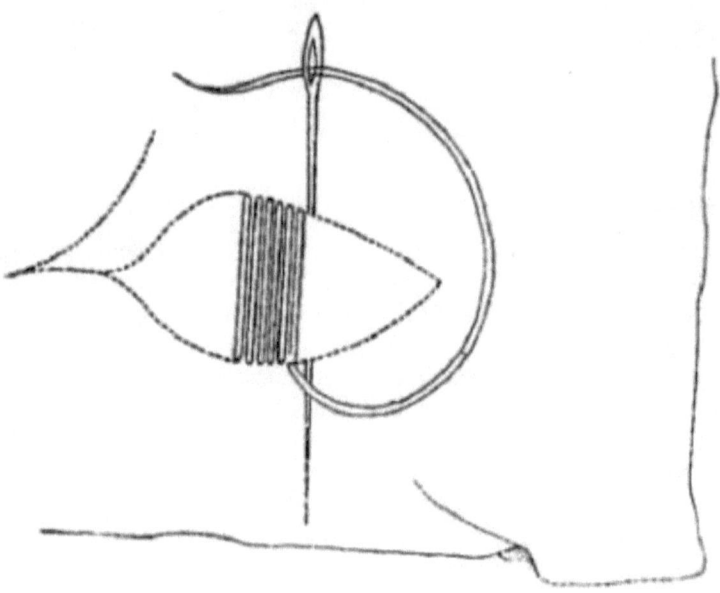

No. 4.—Satin Stitch.

Blanket Stitch is used for working the edges of [Pg 24] tablecovers, mantel valances, blankets, &c., or for edging any other material. It is simply a button-hole stitch, and may be varied in many ways by sloping the stitches alternately to right and left; by working two or three together, and leaving a space between them and the next set; or by working a second row round the edge of the cloth over the first with a different shade of wool.

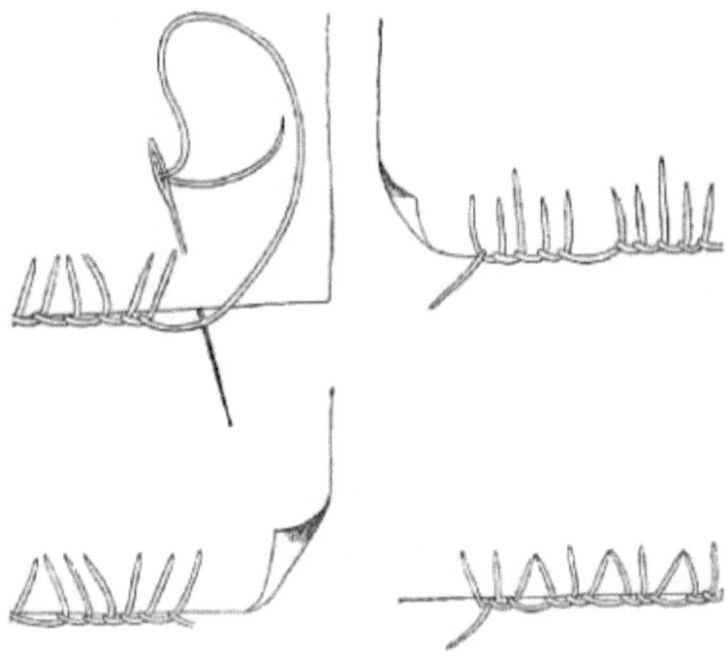

No. 5.—Blanket Stitch.

Knotted Stitch, or **French Knot**, is used for the centres of such flowers as the daisy or wild rose, and sometimes for the anthers of others. The needle is brought up at the exact spot where the knot is to be: the thread is held in the left hand, and twisted once or twice round the needle, the point of which is then passed through the [Pg 25] fabric close to the spot where it came up: the right hand draws it underneath, while the thumb of the left keeps the thread in its place until the knot is secure. The knots are increased in size according to the number of twists round the needle. When properly made, they should look like beads, and lie in perfectly even and regular rows.

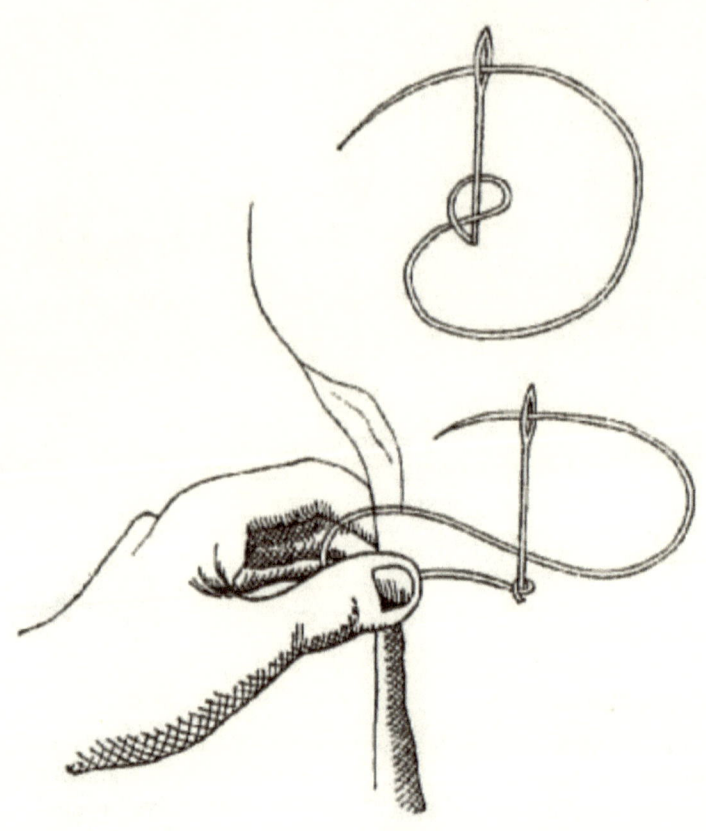

No. 6.—Knotted Stitch, or French Knot.

This stitch is very ancient, and does not seem confined to any country, and the Chinese execute large and elaborate pieces of embroidery in it, introducing beautiful shading. A curious specimen of very fine knotting stitch was exhibited at the Royal School in 1878, probably of French workmanship. It was a portrait of St. Ignatius Loyola, not more than six inches in length, and was entirely executed in knots of such fineness, that without a magnifying glass it was impossible to discover the stitches. This, however, is a *tour de force*, and not quoted as worthy of imitation.

There is one variety of this stitch, in which the thread is twisted a great many times round the needle, so as to form a sort of curl instead of a single knot. This is found in many ancient embroideries, where it is used for the hair of saints and angels in ecclesiastical work.

Knotted stitch was also employed largely in all its forms in the curious and ingenious but ugly style in vogue during the reign of James I., when the landscapes were frequently worked in cross, or feather stitch, while the figures were raised over stuffing, and dressed, as it were, in robes made entirely in point lace, or buttonhole stitches, executed in silk. The foliage of the trees and shrubs which we generally find in these embroidered [Pg 26] pictures, as well as the hair in the figures, were worked in knotted stitches of varying sizes, while the faces were in tent stitch or painted on white silk, and fastened on to the canvas or linen ground.

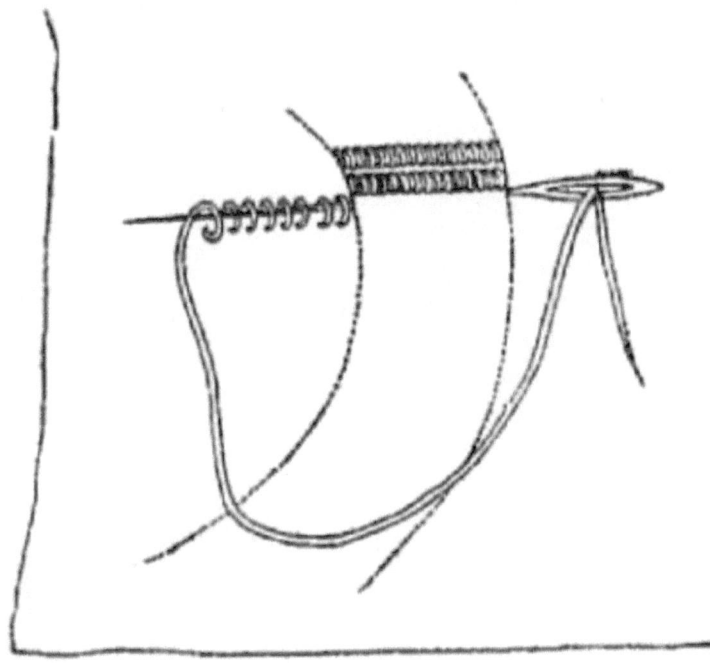

No. 7.—Bullion Knot.

Another variety of knotting, which is still occasionally used, resembles *bullion*, being made into a long roll. A stitch of the length of the intended roll is taken in the material, the point of the needle being brought to the surface again in the same spot from which the thread originally started; the thread is then twisted eight or ten times round the point of the needle, which is drawn out carefully through the tunnel formed by the twists, this being kept in its place by the left thumb. The point of the needle is then inserted once more in the same place as it first entered the material, the long knot or roll being drawn so as to lie evenly between the points of insertion and re-appearance, thus treating the twisted thread as if it were bullion or purl.

[Pg 27] **Chain Stitch** is but little used in embroidery now, although it may sometimes be suitable for lines. It is made by taking a stitch from right to left, and before the needle is drawn out the thread is brought round towards the worker, and under the point of the needle.

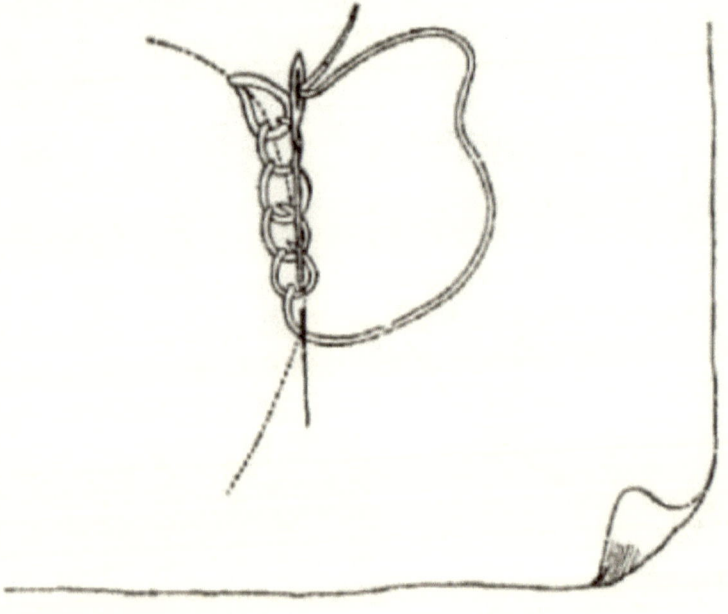

No. 8.—Chain Stitch.

The next stitch is taken from the point of the loop thus formed forwards, and the thread again kept under the point, so that a regular chain is formed on the surface of the material.

This chain stitch was much employed for ground patterns in the beautiful gold-coloured work on linen for dress or furniture which prevailed from the time of James I. to the middle of the eighteenth century. It gave the appearance of quilting when worked on linen in geometrical designs, or in fine and often-repeated arabesques. Examples of it come to us from Germany and Spain, in which the design is embroidered in satin stitch, [Pg 28] or entirely filled in with solid chain stitch, in a uniform gold colour.

Chain stitch resembles *Tambour work*, which we shall describe amongst framework stitches, though it is not at present practised at this School.

Twisted Chain, or Rope stitch.

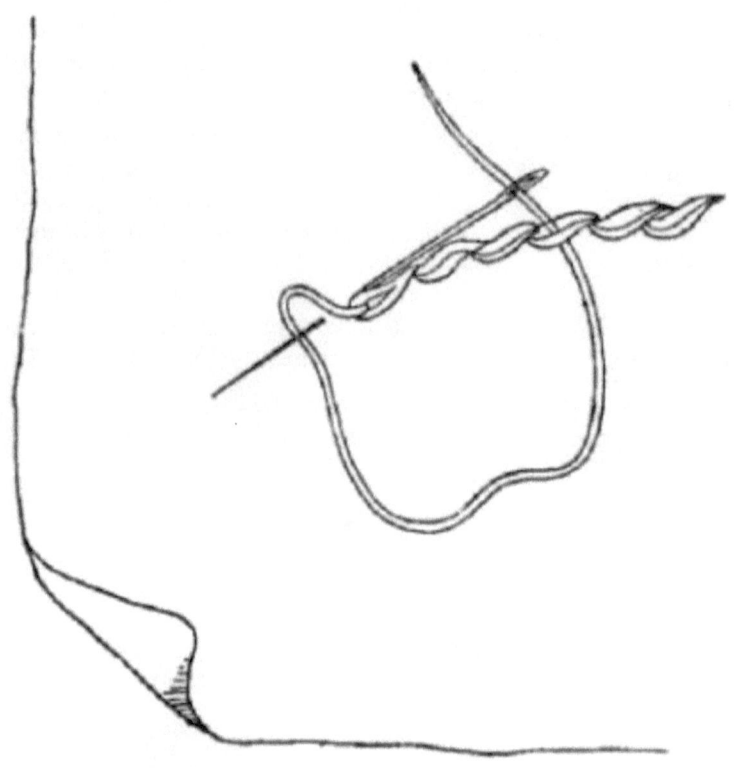

No. 9.—Twisted Chain.

Effective for outlines on coarse materials, such as blankets, carriage rugs, footstools, &c.

It is like an ordinary chain, except that in place of starting the second stitch from the centre of the loop, the needle is taken back to half the distance behind it, and the loop is pushed to one side to allow the needle to enter in a straight line with the former stitch. It is not of much use, except when worked with double crewel [Pg 29] or with tapestry wool; and should then have the appearance of a twisted rope.

Feather Stitch.—Vulgarly called *"long and short stitch," "long stitch"* and sometimes *"embroidery stitch."* We propose to restore to it its ancient title of feather stitch—"*Opus Plumarium,*" so called from its supposed resemblance to the plumage of a bird.

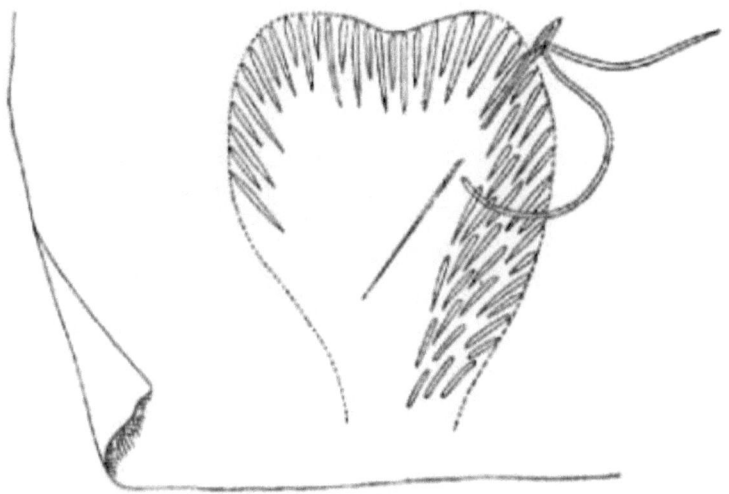

No. 10.—Feather Stitch.

We shall now describe it as used for handwork; and later (at page **37**), as worked in a frame. These two modes differ very little in appearance, as the principle is the same, namely, that the stitches are of varying length, and are worked into and between each other, adapting themselves to the form of the design, but in handwork the needle is kept on the surface of the material.

[Pg 30] Feather Stitch is generally used for embroidering flowers, whether natural or conventional.

In working the petal of a flower (such as we have chosen for our illustration), the outer part is first worked in with stitches which form a close, even edge on the outline, but a broken one towards the centre of the petal, being alternately long and short. These edging stitches resemble satin stitch in so far that the same amount of crewel or silk appears on the under, as on the upper side of the work: they must slope towards the narrow part of the petal.

The next stitches are somewhat like an irregular "stem," inasmuch as they are longer on the surface than on the under side, and are worked in between the uneven lengths of the edging stitches so as to blend with them. The petal is then filled up by other stitches, which start from the centre, and are carried between those already worked.

When the petal is finished, the rows of stitches should be so merged in each other that they cannot be distinguished, and when shading is used, the colours should appear to melt into each other.

In serrated leaves, such as hawthorn or virginia creeper, the edging stitches follow the broken outline of the leaf instead of forming an even outer edge.

It is necessary to master thoroughly this most important stitch, but practice only can make the worker perfect.

The work should always be started by running the thread a little way in front of the embroidery. Knots should never be used except in rare cases, when it is [Pg 31] impossible to avoid them. The thread should always be finished off on the surface of the work, never at the back, where there should be no needless waste of material. No untidy ends or knots should ever appear there; in fact, the wrong side should be quite as neat as the right. It is a mistake to suppose that pasting will ever do away with the evil effects of careless work, or will steady embroidery which has been commenced with knots, and finished with loose ends at the back.

The stitches vary constantly according to their application, and good embroiderers differ in their manner of using them: some preferring to carry the thread back towards the centre of the petal, on the surface of the work, so as to avoid waste of material; others making their stitches as in satin stitch—the same on both sides, but these details may be left to the intelligence and taste of the worker, who should never be afraid of trying experiments, or working out new ideas.

Nor should she ever fear to unpick her work; for only by experiment can she succeed in finding the best combinations, and, one little piece ill done, will be sufficient to spoil her whole embroidery, as no touching-up can afterwards improve it.

We have now named the principal stitches used in hand embroidery, whether to be executed in crewel or silk.

There are, however, numberless other stitches used in crewel embroidery: such as ordinary stitching, like that used in plain needlework, in which many designs were formerly traced on quilted backgrounds—others, again, are many of them lace stitches, or forms of herringbone, [Pg 32] and are used for filling in the foliage of large conventional floriated designs, such as we are accustomed to see in the English crewel work of the sixteenth and seventeenth centuries, on a twilled cotton material, resembling our modern Bolton sheeting.

It would be impossible to describe or even enumerate them all; as varieties may be constantly invented by an ingenious worker to enrich her design, and in lace work there are already 100 named stitches, which occasionally are used in decorative embroidery. Most of these, if required, can be shown as taught at the Royal School of Art-Needlework, and are illustrated by samplers.

[Pg 33]

CHAPTER IV.

FRAMES AND FRAMING.

Before proceeding to describe the various stitches used in frame embroidery, we will say a few words as to the frame itself, the manner of stretching the material in it, and the best and least fatiguing method of working at it.

The essential parts of an embroidery frame are: first, the bars, which have stout webbing nailed along them, and mortice holes at the ends; second, the stretchers, which are usually flat pieces of wood, furnished with holes at the ends to allow of their being fastened by metal pegs into the mortice holes of the bars when the work is stretched.

In some cases the stretchers are fastened into the bars by strong iron screws, which are held by nuts.

FRAMING.

In choosing a frame for a piece of embroidery we must see that the webbing attached to the sides of the bar is long enough to take the work in one direction. Begin by [Pg 34] sewing the edge of the material closely with strong linen thread on to this webbing. If the work is too long to be put into the frame at one time (as in the case of borders for curtains, table-covers, &c.), all but the portion about to be worked should be rolled round one bar of the frame, putting silver paper and a piece of wadding between the material and the wood, so as to prevent its being marked.

The stretchers should then be put in and secured with the metal pegs.

A piece of the webbing having been previously stitched on to the sides of the material, it should now be braced with twine by means of a packing needle, passing the string over the stretchers between each stitch taken in the webbing, and, finally, drawing up the bracing until the material is strained evenly and tightly in the frame. If the fabric is one which stretches easily, the bracings should not be drawn too tightly.

For small pieces of work a deal hand-frame, morticed at the corners, will suffice, and this may be rested on the table before the worker, being held in its position by two heavy leaden weights, covered with leather or baize, in order to prevent them from slipping. It should be raised off the table to a convenient height, thus saving the worker from stooping over her frame, which tires the eyes, and causes the blood to flow to the head.

There is no doubt that a well-made standing-frame is a great convenience, as its position need not be disturbed, and it can be easily covered up and put aside when not in use. It requires, however, to be very well made, and should, if possible, be of oak or mahogany, or it will [Pg 35] warp and get out of order. It must also be well weighted to keep it steady.

For a large piece of work it is necessary to have a long heavy frame with wooden trestles, on which to rest it. The trestles should be made so as to enable the frame to be raised or lowered at will.

A new frame has recently been invented and is sold by the Royal School, which, being made with hinges and small upright pins, holds the ends of the material firmly, so that it can be rolled round and round the bar of the frame without the trouble of sewing it on to the webbing.

When a frame is not in use, care should be taken that it does not become warped from being kept in too dry or too hot a place, as it is then difficult to frame the work satisfactorily.

It will be found useful to have a small basket, lined with holland or silk, fastened to the side of the frame, to hold the silks, thimbles, scissors, &c., needed for the work. Two thimbles should be used, one on each hand, and the best are old silver or gold ones, with all the roughness worn off, or ivory or vulcanite.

The worker ought to wear a large apron with a bib to save her dress, and a pair of linen sleeves to prevent the cuffs from fraying or soiling her work.

Surgeon's bent scissors are useful for frame embroidery, but they are not necessary, as ordinary sharp-pointed scissors will answer every purpose.

When silk, satin, or velvet is not strong enough to bear the strain of framing and embroidering, it must be backed with a fine cotton or linen lining. The "backing" in this case is first framed, as described above, and the velvet or [Pg 36] satin must then be laid on it, and first fastened down with pins; then sewn down with herringbone stitch, taking care that it is kept perfectly even with the thread of the "backing," and not allowed to wrinkle or blister.

It is most important that a worker should learn to use equally both hands, keeping the right hand above the frame till the arm is tired, then letting the left take its place while the right goes below.

A cover should be made large enough to envelop both the upper and under portions of the work, and to be fastened down to the sides, so as to protect it from dust when it is not being used, and during work it should be kept over the portion of the embroidery not actually in hand.

Lastly, a good light should be chosen, so as not to try the eyes.

Many materials can only be embroidered in a frame, and most work is best so done. A greater variety of stitches is possible, and on the stretched flat surface the worker can see the whole picture at once, and judge of the effect of the colours and shading as she carries out the design. It is the difference between drawing on stretched or crumpled paper.

[Pg 37]

CHAPTER V.

STITCHES USED IN FRAME EMBROIDERY.

Feather Stitch.—In framework, as in handwork, we restore the ancient name of *Feather work* or stitch—*Opus Plumarium*. We have already said that it was so-called from its likeness to the plumage of a bird.

This comes from the even lie of the stitches, which fit into and appear to overlap each other, presenting thus a marked contrast to the granulated effect of tent stitches, and the long ridges of the *Opus Anglicum*, having no hard lines as in stem stitch, or flat surfaces as in satin stitch.

Feather stitch, when worked in a frame, is exactly the same as that worked in the hand, except that it is more even and smooth. The needle is taken backwards and forwards through the material in stitches of varying lengths; the next row always fitting into the vacant spaces and projecting beyond them, so as to prepare for the following row.

Every possible gradation of colour can be effected in [Pg 38] this way, and it applies to every form of design—floral or arabesque. Natural flowers have mostly been worked in this stitch.

A skilful embroiderer will be careful not to waste more silk than is absolutely necessary on the back of the work, while, at the same time, she will not sacrifice the artistic effect by being too sparing of her back stitches.

[Pg 39]

"COUCHING," OR LAID EMBROIDERY.

This name is properly applied to all forms of embroidery in which the threads of crewel, silk, or gold are laid on the surface, and stitched on to it by threads coming from the back of the material. Under this head may be classed as varieties the ordinary "laid backgrounds," "diaper couchings," "brick stitch," "basket stitch," and the various forms of stuffed couchings which are found in ancient embroideries. Couching outlines are usually thick strands of double crewel, tapestry wool, filoselle, cord, or narrow ribbon laid down and stitched at regular intervals by threads crossing the couching line at right angles. They are used for coarse outline work, or for finishing the edges of appliqué.

Plain Couching, or "**Laid Embroidery**."—The threads are first laid evenly and straight from side to side of the space to be filled in, whether in the direction of warp or woof depends on the pattern; the needle being passed through to the back, and brought up again not quite [Pg 40] close, but at a sufficient distance to allow of an intermediate stitch being taken backwards; thus the threads would be laid alternately first, third, second, fourth, and so on. This gives a better purchase at each end than if they were laid consecutively in a straight line. If the line slants much, it is not necessary to alternate the rows. When the layer is complete, threads of metal, or of the same or different colour and texture, are laid across at regular intervals, and are fixed down by stitches from the back.

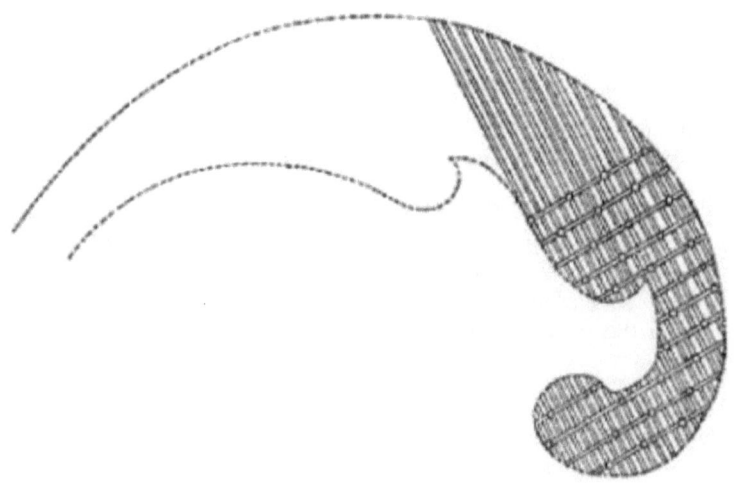

No. 11.—Plain Couching.

The beauty of this work depends upon its regularity.

This kind of embroidery, which we find amongst the old Spanish, Cretan, and Italian specimens, is very useful where broad, flat effects without shading are required; but unless it is very closely stitched down, it is not durable [Pg 41] if there is any risk of its being exposed to rough usage. It is possible to obtain very fine effects of colour in this style of work, as was seen in the old Venetian curtains transferred and copied for Louisa, Lady Ashburton. These were shown at the time of the Exhibition of Ancient Needlework at the School in 1878.

Ancient embroidery can be beautifully restored by grounding in "laid work," instead of transferring it where the ground is frayed, and the work is worthy of preservation. It must be stretched on a new backing, the frayed material carefully cut away, and the new ground couched as we have described.

In other varieties of couching, under which come the many forms of diapering, the threads are "laid" in the same manner as for ordinary couching; but in place of laying couching lines across these, the threads of the first layer are simply stitched down from the back, frequently with threads of another colour.

Net-patterned Couching.—The fastening stitches are placed diagonally instead of at right angles, forming a network, and are kept in place by a cross-stitch at each intersection.

This style of couching was commonly used as a ground in ecclesiastical work of the fourteenth and fifteenth centuries.

Brick Stitch.—The threads are laid down two together, and are stitched across at regular intervals. The next two threads are then placed together by the side, the fastening stitches being taken at the same distance from [Pg 42] each other, but so as to occur exactly between the previous couplings. Thus giving the effect of brickwork.

Diaper Couchings.—By varying the position of the fastening stitches different patterns may be produced, such as diagonal crossings, diamonds, zigzags, curves, &c.

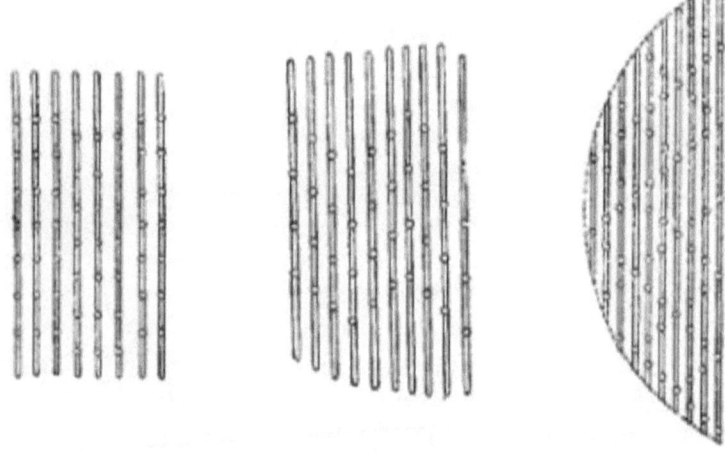

No. 12.—Three Illustrations of Diaper Couchings.

They are properly all gold stitches; but purse silk, thin cord, or even untwisted silk may be used.

A wonderful example of the many varieties of diapering is to be seen in the South Kensington Museum, No. 689. It is modern Belgian work, executed for the Paris Exhibition of 1867. As a specimen of fine and beautiful diapering in gold, this could scarcely be surpassed.

Basket Stitch is one of the richest and most ornamental of these ancient modes of couching. Rows of "stuffing," manufactured in the form of soft cotton cord, are laid [Pg 43] across the pattern and firmly secured. Across these are placed gold threads, two at a time, and these are stitched down over each two rows of stuffing. The two gold threads are turned at the edge of the pattern, and brought back close to the last, and fastened in the same way. Three double rows of gold may be stitched over the same two rows of stuffing.

The next three rows must be treated as brick stitch, and fastened exactly between the previous stitchings, and so on, until the whole space to be worked is closely covered with what appears to be a golden wicker-work.

Strong silk must be used for the stitching.

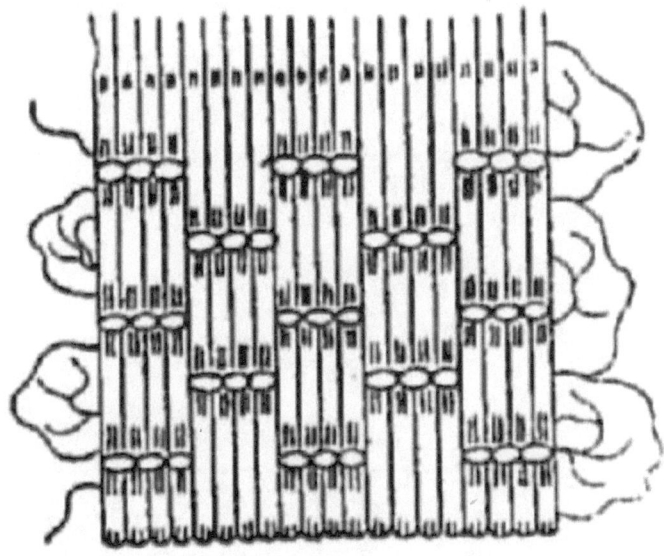

No. 13.—Basket Stitch.

The Spanish School of Embroidery has always been famed for its excellence in this style, and has never lost the art. The "Embroiderers of the King," as they are called, still turn out splendid specimens of this heavy and elaborate work, which are used for the gorgeous trappings of the horses of the nobility on gala days and state occasions.

A beautiful specimen was exhibited at the Royal School of Art-Needlework, in 1878, by the Countess Brownlow, of an altar-hanging, entirely worked in basket [Pg 44] stitch, in gold on white satin, and a modern example is still to be seen at the School in a large counterpane, which was worked for the Philadelphia Exhibition from an ancient one also belonging to Lady Brownlow.

The Spanish embroiderers used these forms of couching over stuffing with coloured silks as well as gold, and produced wonderfully rich effects. One quilt exhibited by Mrs. Alfred Morrison in 1878 was a marvel of colouring and workmanship.

Basket stitch is mostly used now for church embroidery, or for small articles of luxury, such as ornamental pockets, caskets, &c.

Diapering is generally employed in the drapery of small figures, and in ecclesiastical work.

Many fabrics are manufactured in imitation of the older diapered backgrounds, and are largely used to replace them. Among these are the material known as silk brocatine, and several kinds of cloth of gold mentioned in our list of materials.

[Pg 45]

CUSHION STITCHES.

Cushion Stitch—the ancient *Opus Pulvinarium* of the Middle Ages, likewise called "Cross Stitch"—may lay claim to be one of the most ancient known in embroidery. There have been many varieties, but the principle is the same in all. It is worked on and through canvas, of which the threads, as in tapestry, regulate the stitches.

After six centuries of popularity it finally died out within the last few years as "Berlin wool work;" but will doubtless be revived again in some form after a time, as being well fitted for covering furniture on account of its firmness and durability.

In Germany and Russia it is still much used for embroidering conventional designs on linen; and the beautiful Cretan and Persian work of which so much has lately been in the market, is executed in this style.

Tent Stitch may be placed first under this class, in [Pg 46] which the thread coming from beneath is carried over a single cross of the warp and woof of the canvas.

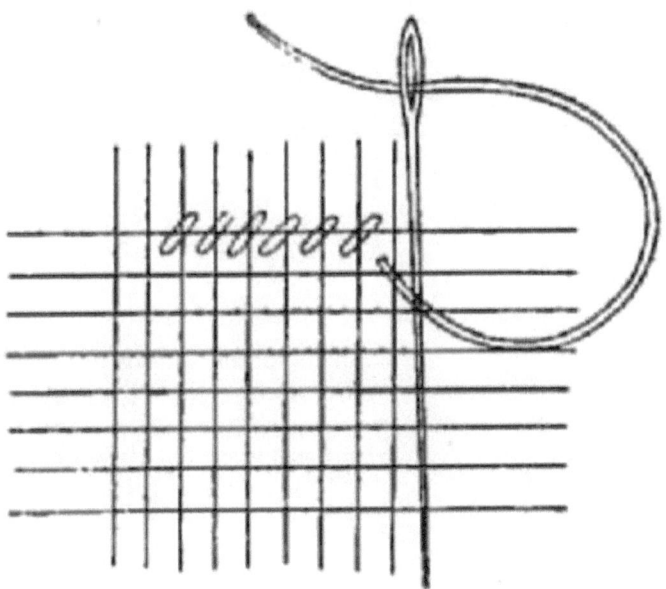

No. 14.—Tent Stitch.

Simple Cross Stitch.—The worsted or silk is brought up again to the surface, one thread to the left of the spot where the needle was inserted, and is crossed over the first or "tent" stitch, forming a regular and even cross on the surface.

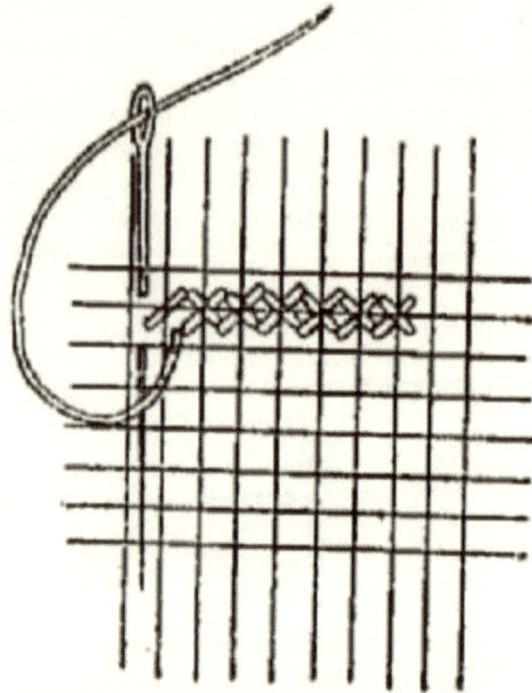

No. 15.—Simple Cross Stitch.

Persian Cross Stitch.—The peculiarity of this stitch is that in the first instance the silk or worsted is carried [Pg 47] across two threads of the canvas ground, and is brought up in the intermediate space. It is then crossed over the latter half of the original stitch, and a fresh start is made.

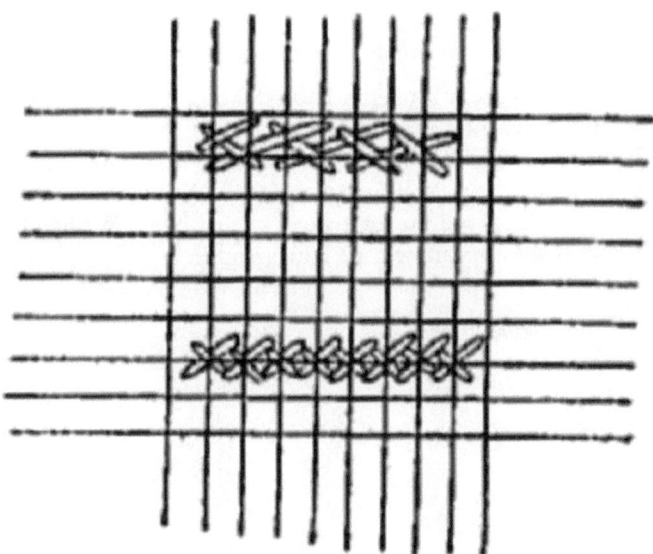

No. 16.—Persian Cross Stitch.

Much of the beauty of Persian embroidery is produced by the irregularity of the crossing; the stitches being taken in masses, in any direction that seems most suitable to the design in hand, instead of being placed in regular rows, with the stitches all sloping in one direction, as is the case with the modern "Berlin work," this, with the happy choice of colours for which the Persians are so justly famous, produces a singular richness of effect.

Allied to these canvas stitches and having their origin in them, are the numerous forms of groundings, which are now worked on coarse linens, or in fact on any fabric; and have sometimes, although incorrectly, been called darning stitches, probably from their resemblance to the patterns which are found on samplers, for darning stockings, old table linen, &c. &c. Almost any pattern can be produced in this style of embroidery, simply by varying the relative length of the stitches.

Following the nomenclature of the committee which named and catalogued the specimens of ancient needlework exhibited in the

South Kensington Museum in 1872, we have classed all the varieties of these grounding stitches under the name of Cushion stitch.

Cushion Stitches are taken as in laid embroidery, so as to leave all the silk and crewel on the surface, and only a single thread of the ground is taken up; but in place of lying in long lines, from end to end of the material, they [Pg 48] are of even length, and are taken in a pattern, such as a waved line or zigzag; so that when finished the ground presents the appearance of a woven fabric.

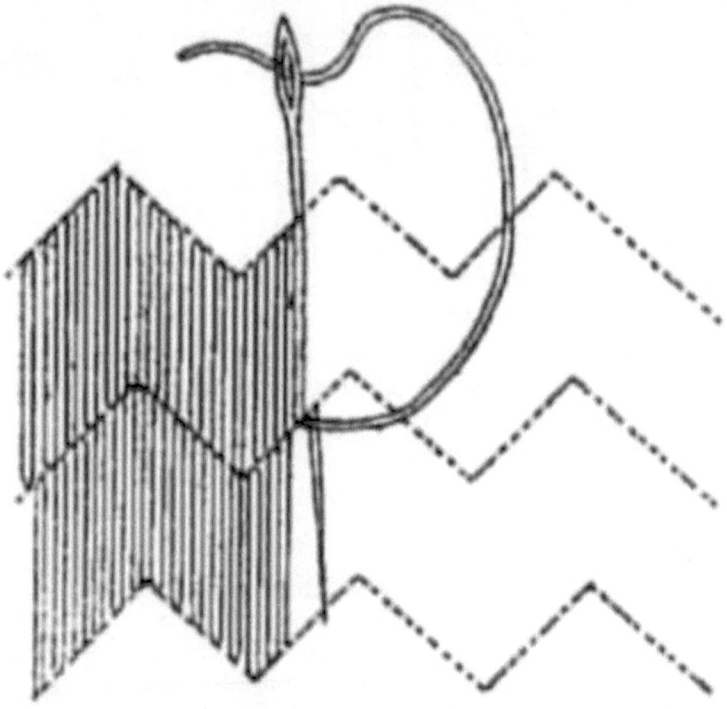

No. 17.—Cushion Stitch.

We give an illustration of one variety of cushion stitch, which may either be worked as described here, or in the hand, as in the woodcut.

A good modern example of this background was exhibited in the School, on a bed-hanging, worked for the Honourable Mrs. Percy

Wyndham, from a design by Mr. W. Morris. In the Exhibition of Ancient Needlework last year were many beautiful specimens: notably one enormous wall-hanging of Italian seventeenth-century work, lent by Earl Spencer. Many of the fabrics known as "Tapestries" are woven imitations of these grounds, and carry embroidery so perfectly, that on the whole, except for small pieces, it seems a waste of hand-labour to work them in, as the effect is not very far removed from that of woven material, while the expense is, of course, very much greater.

[Pg 49] The ancient specimens of this stitch are worked on a coarse canvas, differing greatly from that which was recently used for Berlin wool work.

It cannot now be obtained except by having it especially made to order. It has been replaced by a coarse hand-woven linen for the use of the School, but the ancient canvas is vastly superior, as its looseness makes it easier for the worker to keep her stitches in regular lines.

In some ancient specimens the design is worked in feather stitch, and the whole ground in cushion stitch. In others the design is in fine cross or tent stitch. There are several very beautiful examples of this kind of embroidery in the South Kensington Museum—Italian, of the seventeenth century.

A variety of cushion stitch, which we frequently see in old Italian embroideries, was taught in the Royal School of Art-Needlework by Miss Burden, and used under her direction in working flesh in some large figures designed by Mr. Walter Crane for wall decoration, and exhibited at the Centennial Exhibition at Philadelphia. The stitches are kept of one uniform length across the design. The next row is started from half the depth of the preceding stitch and kept of the same length throughout. Its beauty consists in its perfect regularity. If worked in the hand, the needle is brought back underneath the material as in satin stitch; but in the frame all the silk or worsted can be worked on the surface, with the exception of the small fastening stitches.

The effect when finished is that of a woven fabric.

It is really more suitable in its original character of a ground stitch than for working flesh. We have given an [Pg 50] illustration of it, because we are so frequently asked to describe "Burden stitch."

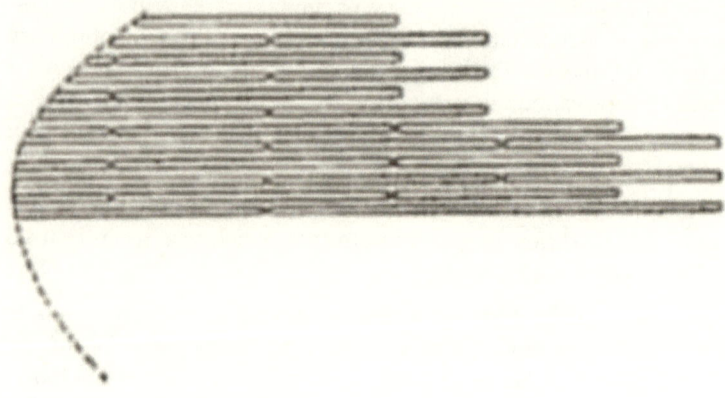

No. 18.—"Burden" Stitch.

This form of cushion stitch worked extremely fine has been used for flesh in very ancient embroideries, even before the introduction of the *Opus Anglicanum*, and is found in the works of the Flemish, German, Italian, and French schools of the fourteenth and fifteenth centuries.

It seems to have been worked in a frame on fine canvas, or on a fabric of very even threads, and the stitches so taken that the same amount of silk appears on the back as on the surface of the embroidery.

In a toilet cover of ancient Spanish work recently added to the South Kensington Museum, the design is entirely embroidered in varieties of *cushion stitch* in black floss silk upon a white linen ground. It is, however, extremely rare to see this stitch used in any other way than as a ground, except in actual canvas work; in which we often see varieties of it used to fill in portions of the design, while another stitch will be devoted entirely to the grounding.

These stitches were often executed on an open net.

[Pg 51] **Stem Stitch** is used in frame embroidery, and does not differ in any way from that described at page **20** , under "handwork," except that the needle is of course worked through the material with both hands, as is the case in all frame work.

The same may be said of "split stitch;" but this is more frequently (because more easily) worked in a frame than done in the hand.

Japanese Stitch is a modification of stem, but its peculiarity consists in the worker taking very long stitches, and then bringing the needle back to within a short distance of the first starting-place; so that they may be in even parallel lines, advancing by gradation from left to right. It is principally used for working water or ground in a landscape.

No. 19.

Tambour Work has fallen into disuse, but was greatly admired when our grandmothers in the last century sprigged Indian muslins or silks with coloured flowers for dresses, and copied or adapted Indian designs on fine linen coverlets. These were very refined, but no more effective than a good chintz. There are exquisite specimens of the stitch to be seen in most English homes, and in France it was in vogue in the days of Marie Antoinette. Its use is now almost confined to the [Pg 52] manufacture of what is known as Irish or Limerick lace, which is made on net in the old tambour frames, and with a tambour or crochet hook. The frame is formed of two rings of wood or iron, made to fit loosely one within the other. Both rings are covered with baize or flannel wound round them till the inner one can only just be passed through the outer. The fabric to be embroidered is placed over the smaller hoop, and the other is pressed down over it and firmly fixed with a screw. A small wooden frame of this description is universally used in Ireland for white embroidery on linen or muslin. In tambour work the thread is kept below the frame and guided by the left hand, while the hook or crochet needle is passed from the surface through the fabric, and brings up a loop of the thread through the preceding stitch, and the needle again inserted, forming thus a close chain on the surface of the work.

The difficulty of working chain stitch in a frame probably led to the introduction of a hook for this class of embroidery.

Perhaps we ought not to omit all mention of the **Opus Anglicum** or *Anglicanum* (English work), though it is strictly ecclesiastical, and therefore does not enter into our province.

Dr. Rock [1] and other authorities agree in thinking that the distinctive feature of this style, which was introduced about the end of the thirteenth century, was a new way of working the flesh in subjects containing figures.

[Pg 53] Instead of the faces being worked in rows of straight stitches (like that described as Burden stitch on page **50**) as we see in the old Flemish, German, and Italian work of the same period, the English embroiderers invented a new stitch, which they commenced in the centre of the cheek and worked round and round—gradually letting the lines fall into outer circles of ordinary feather stitch.

Having thus prepared an elastic surface, they proceeded to model the forms and make lights and shadows by pressing the work into hollows, with small heated metal balls, the work being probably damped as a preparation for this process. So skilfully did they carry out their intention, that the effect is still the same after the lapse of five centuries. We must unwillingly add that, though much appreciated in the thirteenth century, the effect is rather curious and quaint than beautiful.

The Syon cope in the Kensington Museum, of the thirteenth century, is a fine specimen of this attempt to give the effect of bas-relief to the sacred subjects depicted. The whole cope shows how various were the stitches worked at that period. On examination with a microscope, the flesh stitch appears to be merely a fine split stitch worked spirally, as we now work fruit.

FOOTNOTE:

[1] See Dr. Rock's preface to his "Descriptive Catalogue of Textile Fabrics" in the Kensington Museum.

[Pg 54]

CUT WORK OR APPLIQUÉ.

Decorative cut work is of infinite variety, but may be divided into two groups, "inlaid appliqué" and "onlaid appliqué."

"*Inlaid*" appliqué consists in tracing the same pattern on two different fabrics, say a gold cloth and a crimson velvet; then cutting both out carefully, and inlaying the gold flowers into the crimson velvet ground, and the crimson flowers into the gold ground.

This kind of work may be seen constantly in Italian rooms of the seventeenth century, and the alternate breadths of crimson and gold give a very fine effect as of pilasters, and in general are enriched by a valance applied at the top, and a plain border at the bottom.

The *inlaid* part is sewn down with thread, and covered with cord or couchings of floss silk. Sometimes narrow ribbons or fine strips of cut silk are stitched over the edges to keep them down flat.

"*Onlaid* appliqué" is done by cutting out the pattern in one or many coloured materials, and laying it down on an intact ground of another material. Parts are often shaded with a brush, high lights and details worked in with stitches of silk, and sometimes whole flowers or figures are embroidered, cut out, and couched [Pg 55] down. This sort of work is extremely amusing, and gives scope to much play of fancy and ingenuity, and when artistically composed it is sometimes very beautiful.

Another style of "onlaid appliqué" is only worked in solid outlines, laid down in ribbon or cord, sometimes in both. This was much in vogue in the time of Queen Anne, and for a hundred years after.

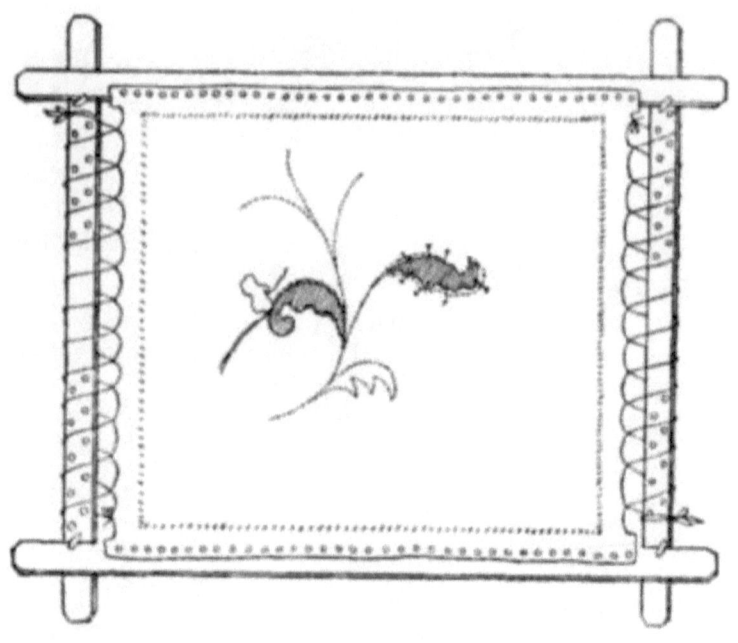

No. 20.

The ribbon, very soft and thick, sometimes figured, sometimes plain, was manufactured with a stout thread on each side, which could be drawn, and so regulate the ribbon and enable it to follow the flow of the pattern.

The German, French, and Italians often enriched this style of work with a flower, embroidered and applied thrown in here and there. Very small fringes also were introduced into the pattern, or arabesqued.

[Pg 56] "Cut work," like the appellation "Feather stitch," has a totally different meaning when it is given to white embroidery, and it has nothing to do with appliqué, but takes its name from the fact that the pattern is mostly cut or punched out, and then edged with button-hole or plain overlaid stitch.

In working appliqué it is best, although not absolutely necessary, to have the design traced on the material to be used as a ground, which must then be framed as for ordinary embroidery. A copy of the design must be made on tracing-paper, and the outlines carefully pricked out with a needle or pin, laying the paper on several folds of flannel or cloth for greater convenience in pricking.

A pad, made of a long strip of flannel about four inches wide, rolled very tightly, must be made ready, and some pounce made of about equal quantities of finely powdered charcoal and pipe-clay. The leaf or scroll which is wanted for the work must now be selected, and the pricked design laid face downwards on the fabric which is to be applied. The flannel pad must be dipped in the pounce and rubbed well into the outlines of the pricked design, which must be held firmly in its place with the left hand. On lifting the tracing-paper, the design will be found to be marked out on the material distinctly enough for it to be cut out with a sharp pair of scissors. The pounce can afterwards be dusted off.

The leaf or scroll having been thus cut out must be fastened in its place on the design with small pins, and then carefully sewn down. The edges are then finished off by stitches of embroidery or by a couching line (*see* page **39**). The stems are frequently worked in with stem [Pg 57] stitching or couching, and the leaves enriched by large veinings of crewel or silk work, or in conventional designs, with some of the many varieties of herringboning.

Gold Embroidery on velvet or satin grounds requires to be worked on a strong even linen, and then cut out and applied in the same manner as ordinary appliqué. Where a particularly rich and raised effect is required any embroidery may be treated in this manner. It is of course more troublesome, but quite repays the labour spent upon it by the increased beauty of the work.

The transfer of old embroideries on to a new ground is usually done by appliqué, although we have already described a better process at page **39** .

In transferring old needlework it is necessary to cut away the ground close to the edge of the embroidery. It is then placed on the new material, which has been previously framed, and the outline tacked down. The best way of finishing is then to work in the edges with silks *dyed exactly to match* the colours in the old work. If properly done, it is impossible to discover which are old and which new stitches, and only by examining the back, that the work has been transferred at all.

We used the words "*dyed to match*" advisedly, as it is impossible otherwise to procure new silks which will correspond with the old.

Embroidery transferred in this manner is as good as it was in its first days, and in many cases is much better, for time often has the same mellowing and beautifying effect in embroideries as in paintings.

A less expensive, but also a much less charming, [Pg 58] method is to edge the old embroidery after applying it to the new ground with a cord or line of couching.

With this treatment it is, however, always easy to perceive that the work has been transferred.

For almost all kinds of appliqué it is necessary to back the material; and it is done in this manner: —

A piece of thin cotton or linen fabric is stretched tightly on to a board with tacks or drawing-pins. It is then covered smoothly, and completely, with paste. The wrong side of the velvet, satin, serge, or whatever is to be used in the work, is then pressed firmly down on the pasted surface with the hands, and then left to dry.

In giving the foregoing account of the most typical stitches, we hope we have succeeded in showing the principle on which each should be worked. They form the basis of all embroidery, and their numerous modifications cannot be fully discussed in the limit we have prescribed to ourselves. It is sufficient to observe that the instruction we have tried to impart is that which it is absolutely necessary for the needleworker to master thoroughly before she attempts to cope with the artistic element of her work. That it is a creative art is undoubted, for no two pieces of embroidery are alike unless executed by the same hand, and from the same design.

For the advanced artist there is a store of instruction in the fine collection at South Kensington, which, seen by the light of Dr. Rock's invaluable "Catalogue of Textile Fabrics," is an education in itself, of which the ethnological as well as the artistic interest cannot be over-estimated, and it is within the reach of all who can find time to bestow upon it.

[Pg 59]

STRETCHING AND FINISHING.

Always avoid using an iron to embroidery. It flattens the work, and is apt to injure the colour. For embroidery on linen, unless very badly done, it will be found quite sufficient to stretch the work as tightly as possible with white tacks or drawing-pins on a clean board, and damp it evenly with a sponge. Leave it until quite dry, and then unfasten it, and, if necessary, comb out the fringe. If it is new work, it should not be fringed until after it has been stretched.

For crewel work on cloth or serge, it is sometimes necessary to rub a little shoemaker's paste on to the back of the embroidery, while it is tightly stretched. When pasting can be avoided, it is always better to do without it; but it serves to steady the work in some cases, and makes it wear better. Unless it is absolutely necessary, it is better not to paste the back of screen panels, whatever may be the materials on which they are worked; but more especially satin or velvet, as it interferes with the straining of the work by the cabinet-maker.

We give a recipe for Embroidery Paste, which is said to be excellent:—Three and a half spoonfuls of flour, and as much powdered resin as will lie on a half-penny. Mix these well and smoothly with half a pint of water, and pour it into an iron saucepan. Put in one teaspoonful of essence of cloves, and go on stirring till it [Pg 60] boils. Let it boil for five minutes, and turn it into a gallipot to cool.

N.B.—Let the gallipot have in it a muslin bag: the water can then be drained out from time to time, and the paste will be much better.

CLEANING.

Good crewels will always wash or clean without injury; but the cheap and inferior worsteds will not do so. Ordinary crewel work on linen may be washed at home, by plunging it into a lather made by water in which bran has been boiled, or even with simple soap-suds, so long as no soda or washing-powder is used. It should be carefully rinsed without wringing, and hung up to dry. When almost dry, it may be stretched out with drawing-pins on a board, and will not require ironing.

Embroidery on cloth or serge may often be cleaned with benzoline, applied with a piece of clean flannel; but in any case, where a piece of work is much soiled, or in the case of fine d'oyleys, it is safer to send it to the cleaner's.

Messrs. Pullar and Son, Perth Dye Works, are very successful in cleaning all kinds of embroidery without injuring it.

In many cases it may be well dyed – the silk in which the design is worked always showing a different shade from the ground.

[Pg 61]

APPENDIX.

DESIGNS FOR EMBROIDERY.

[Pg 62]

DESCRIPTION OF THE PLATES.

No. 1.—Design for Wall-Panel. By Mr. E. Burne-Jones.

Worked in outline on neutral-tinted hand-woven linen in brown crewel. This style of embroidery is very suitable for internal decoration, where a good broad effect is required without a large amount of labour. A frieze or dado, or complete panelling of a room, may be worked in this way at a comparatively small cost.

No. 2.—Design for Wall or Screen Panel. By Mr. Walter Crane. Representing the Four Elements.

Embroidered in crewels on a silk ground of dead gold colour partly outlined.

No. 3.—Design for Quilt or Table Cover. By Mr. George Aitchison.

A border of sunflowers and pomegranates, with powderings of the same for the centre.

This has been embroidered on cream-coloured satin de chine in solid crewel work, with charming effect, both for a counterpane and curtains.

No. 4.—Design for Wall Panelling or Curtains. By Mr. Fairfax Wade.

To be worked in outline and solid embroidery, in silk or filoselle, on satin de chine.

[Pg 63] No. 5.—Design for Quilt or Couvre-Pied. By Mr. Fairfax Wade. To introduce squares of Greek or guipure lace.

Worked in golden shades of silk on linen, lined with silk of the same colour. The embroidery is partly solid and partly outline, very fine and delicate.

No. 6.—Design for Sofa-back Cover. By Mr. W. Morris.

Worked on hand-woven linen in two shades of gold-coloured silks. Outline.

No. 7.—Design for Sofa-back Cover or Piano Panel. By Mr. George Aitchison.

Worked in two shades of blue silk on hand-woven linen or satin de chine.

No. 8.—Design for Appliqué. By Mr. Fairfax Wade.

Nos. 9 and 10.—Designs for Chair-seats or Cushions. By Miss Jekyll. Periwinkle and Iris.

No. 11.—Design for Border. By Miss Webster. To be worked in outline in silk or crewel.

No. 12.—Design for Border for Curtain or Table Cover. Designed by Miss Burnside, of the R.S.A.N.

No. 13.—Table Border. Designed by Mr. Fairfax Wade. Conventional Buttercup. To be worked either solid or in outline.

[Pg 64] No. 14.—Table Border. Designed by Mr. Walter Crane. For solid embroidery in crewel or silk.

No. 15.—Table Border. Designed by Mr. Walter Crane. For solid embroidery in crewel or silk.

No. 16.—Border. Designed by Miss Mary Herbert, R.S.A.N. For crewel or silk embroidery, either in outline or solid.

No. 17.—Two Panels. Designed by Rev. Selwyn Image. Representing Juno and Minerva.

No. 18.—Two Panels. Designed by Rev. Selwyn Image. Representing Venus and Proserpine. To be worked in outline on linen, as No. 1, or in coloured silks on a groundwork of satin de chine.

No. 19.—Wall Hanging. Designed by Mr. W. Morris. To be worked on linen in outline.

No. 20.—Wall Hanging. Designed by Mr. W. Morris. Worked on linen. Background in Silk Cushion Stitch.

No. 21.—Border for Appliqué. Copied from Ancient Italian work.

No. 22.—Italian Design. A Specimen. Showing the application of transposed Appliqué.

[Pg 65]

1. DESIGN FOR WALL PANEL.
By E. Burne-Jones.

[Pg 67]

2. DESIGN FOR WALL PANEL.
By Walter Crane.

[Pg 69]

3. DESIGN FOR A QUILT OR TABLE COVER.
By George Aitchison.
Vincent Brooks Day & Son, Lith.

[Pg 71]

See larger image
4. DESIGN FOR WALL PANEL OR CURTAIN.
By Fairfax Wade.

[Pg 73]

5. DESIGN FOR A QUILT OR COUVRE-PIED.
By Fairfax Wade.

[Pg 75]

See larger image
6. DESIGN FOR A SOFA-BACK COVER.
By William Morris.

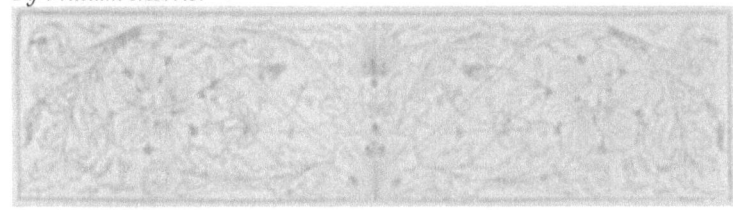

See larger image
7. DESIGN FOR A SOFA-BACK COVER OR PIANO PANEL.
By George Aitchison.
Vincent Brooks Day & Son, Lith.

[Pg 77]

See larger image
8. DESIGN FOR APPLIQUÉ.
By Fairfax Wade.

[Pg 79]

See larger image
DESIGNS FOR CHAIR-SEATS OR CUSHIONS. (9. PERIWINKLE 10. IRIS.)
By Miss Jekyll.
Vincent Brooks Day & Son, Lith.

[Pg 81]

See larger image

11. DESIGN FOR A BORDER.
By Miss Webster.
Vincent Brooks Day & Son, Lith.

12. DESIGN FOR A BORDER FOR A CURTAIN OR TABLE COVER.
By Miss Burnside.

See larger image

See larger image

DESIGNS FOR TABLE BORDERS.
No. 13 by Fairfax Wade; 14 and 15 by Walter Crane; 16 by Mary Herbert.
Vincent Brooks Day & Son, Lith.

[Pg 85]

17. TWO DESIGNS FOR WALL PANELS—"JUNO" AND "MINERVA."
By the Rev. Selwyn Image.

[Pg 87]

18. TWO DESIGNS FOR WALL PANELS—"VENUS" AND "PROSERPINE."
By the Rev. Selwyn Image.

[Pg 89]

See larger image
19. DESIGN FOR WALL-HANGING.
By William Morris.

[Pg 91]

20. DESIGN FOR WALL-HANGING.
By William Morris.
Vincent Brooks Day & Son, Lith.

[Pg 93]

21. DESIGN FOR BORDER FOR APPLIQUÉ.
From Ancient Italian Work.
Vincent Brooks Day & Son, Lith.

[Pg 95]

22. ITALIAN DESIGN.
Showing the application of transposed Appliqué.
Vincent Brooks Day & Son, Lith.

[Pg 97]

Royal School of Art-Needlework.

Incorporated under "The Companies' Acts, 1862 and 1867," by licence of the Board of Trade, granted under 30 and 31 Vic., c. 131, sec. 23.
Share Capital, £10,000, in 1000 Shares of £10 each. Debenture Capital, £10,000, to be issued in Debentures of £50 each.

Patrons.

Her Majesty the Queen.
H.R.H. The Prince of Wales.
H.R.H. The Princess of Wales.

President.

H.R.H. The Princess Christian of Schleswig-Holstein.
Princess of Great Britain and Ireland.

Vice-President.

The Lady Marian Alford.

Managing Committee.

The Countess Spencer.
The Countess Cowper.
The Countess Brownlow.
The Viscountess Downe.
The Lady Sarah Spencer.
The Hon. Lady Welby Gregory.
The Hon. Mrs. Percy Wyndham.
Mrs. Edward Baring.
(With power to add to their number.)

Honorary Members of the Managing Committee.

The Lady Charlotte Schreiber.
The Hon. Lady Hamilton-Gordon.
The Lady Fitzhardinge.
The Hon. Mrs. Stuart Wortley.

[Pg 98]

Finance Committee.

The Duke Of Westminster, K.G.
The Earl Brownlow.
The Lord Sudeley.
Sir Coutts Lindsay, Bart.
The Right Hon. Sir William Henry Gregory, K.C.M.G.
Michael Biddulph, Esq., M.P.
Edmund Oldfield, Esq.

Bankers.

London and County Bank, Albert Gate Branch.

Solicitors.

Messrs. Trinders & Curtis-Hayward, 4, Bishopsgate Street Within, E.C.

Offices.

EXHIBITION ROAD, SOUTH KENSINGTON.

PROSPECTUS.

The School was founded in 1872, under the Presidency of H.R.H. the Princess Christian of Schleswig-Holstein, for the twofold purpose of supplying suitable employment for Gentlewomen and restoring Ornamental Needlework to the high place it once held among the decorative arts.

It was first established, under the title of School of Art-Needlework, in Sloane Street; but in 1875 was removed to the present premises in the Exhibition Road, and Her Majesty the Queen was graciously pleased to grant to it the prefix of "Royal."

The Royal School of Art-Needlework exhibited at the Centennial Exhibition of Philadelphia, 1876, and received a Certificate of Award—medals not being granted to institutions or corporate bodies. A Silver Medal was also granted by the Jurors of the International Exhibition, Paris, 1878, for embroideries exhibited there.

The result of seven years' experience of the working of the School has shown that the objects for which it was formed are appreciated by the public, and has justified its establishment on a permanent basis. This has accordingly been effected under a special licence from the Board of [Pg 99] Trade, granted under authority of an Act of Parliament which authorizes the incorporation of associations *not* constituted for purposes of profit.

The ultimate profits of the Association, after payment of all Debentures, are to be applied to such charitable or other purposes as the Association may from time to time determine, not being inconsistent with the provisions of the Memorandum of Association, which require that the Shareholders shall not take any personal profit out of the Association.

The government of the School is vested in:

First. — A President, Vice-President, and General Council.

Second. — A Managing Committee to be selected from the General Council, except as to Honorary Members to be nominated by the Managing Committee.

Third. — A Finance Committee, of whom a majority are to be elected by the Shareholders, and the remainder nominated by the Managing Committee. The sanction of this Committee is required for all expenditure.

Agencies have now been opened in Liverpool, Manchester, Leeds, Norwich, Birmingham and Glasgow; and a member of the staff has been sent out to take charge of the School of Art-Needlework in Philadelphia.

The Show Rooms are open from 10 a.m. to 6 p.m. in Summer, and to 5 p.m. in Winter, and close on Saturdays at 2 p.m.

All letters must be addressed "The Secretary."

Lists of designs, prices of prepared and finished work, terms for lessons, and addresses of Provincial Agents, may be obtained by writing to the Secretary.

A Branch School for Scotland has now been opened in Glasgow. Show Rooms at 108, St. Vincent Street.

[Pg 100]

ROYAL SCHOOL OF ART-NEEDLEWORK.
EXHIBITION ROAD, SOUTH KENSINGTON.

PREPARED WORK.

Work can be obtained from the Royal School of Art-Needlework having a design traced, a portion of the embroidery commenced, and sufficient materials for finishing. Ladies' own materials will be traced and prepared for working if desired. Dresses must be cut out and tacked together before being sent to the School, and lines marked on the material to show where the design is to be placed.

When an order for prepared work is executed exactly by the directions given, or when the selection of Design or Colouring is left to the School, *the work cannot be exchanged or taken back.*

The materials supplied with the work are considered more than sufficient to finish it, and if more are required afterwards they must be purchased separately.

A few specimen prices are quoted, but *no estimates can be given for prepared work*, except in cases of large orders where a great quantity of material is supplied.

[Pg 101] *All Designs supplied are Copyright of the Royal School of Art-Needlework, and must not be made use of for purposes of sale.*

Designs on paper are not supplied under any circumstances, nor can work be sent out on approbation.

All work supplied is stamped with the monogram of the Royal School of Art-Needlework, as above, in addition to the letters P. W.

N.B.—*An extra charge is made for all designs not ordinarily used for Prepared Work.*

APPROXIMATE PRICES OF PREPARED WORK AND MATERIALS.

Table Covers, on Diagonal, from £1 1s. to £5 5s.

" " Serge " 18s. to £3 3s.

Linen Table Covers, yard square, 14s. 6d. to £1 10s.

Chair Back Covers, Linen, 7s. 6d. to £1 1s.

Borders, on Linen, suitable for Table Covers or Dresses, from 5s. per yard.

Borders, on Serge or Diagonal, suitable for Table Covers or Dresses, from 7s. per yard.

Borders, on Serge or Diagonal, suitable for Curtains, Chimney Valances, &c., from 13s. per yard.

N.B.—*If several yards are ordered of one pattern the price is lower.*

Banner Screens, Linen (various), 8s. 6d. to 15s. 6d.

" " Diagonal, 12s. 6d. to £2 2s.

Babies' Blankets, from 14s. 6d.

Bath Blankets, yard square, 17s. 6d.; yard and a half square, 26s.

Children's Dress, from 18s. to £1 10s.

Tennis Aprons, from £1 1s.

Cushions, Linen, 7s. 6d. to 12s 6d.; on Diagonal, &c., 10s. 6d. to £1 1s.

Toilet Mats or D'Oyley, 8 inches square, from £1 6s. to £3 3s. per dozen.

Folding Screens, on Sailcloth, £1 1s. to £1 10s. per panel.

[Pg 102]

CREWELS.

Crewels are sold at the rate of 8d. per ounce skein, or in quarter-pound bundles, containing not more than four shades, at 2s. In quarter-pound bundles, containing selected colours, at 3s.

Embroidery Silks, at 6s. 6d. per ounce reel, and 3s. 3d. per half-ounce reel of one shade; or at 8s. per ounce of selected colours.

Filoselle, 3s. 6d. per ounce.

Needles, 9d. per packet.

Materials, suitable for embroidery, such as Homespuns, Fancy Linens, Serge, Diagonal, Utrecht Velvet, Satin de Chine, &c. &c., may be purchased at the School.

NOT LESS THAN ONE YARD SOLD.

[Pg 103]

LIST OF DESIGNS.

CHAIR BACKS.

Honeysuckle, Bramble, Poppy, Passion Flower, Taxonia, Wild Rose, Apple Blossom, Orange with Flowers, Virginia Creeper, Fish and Bulrushes, Winter Cherry, Corn Flower, Hops, Carnations, Cherry, Daisy Powdered, Primrose Powdered, Faust Motto, Iris Seed, Japanese, Jessamine, Lantern Plant, Periwinkle, Potato, Zynia, Tiger Lily, Geranium, Burrage, Corncockle, Hawthorn, Daffodil, Iris, Love-in-a-Mist, &c. &c., with many conventional designs.

NARROW BORDERS.
SUITABLE FOR DRESSES OR TABLE COVERS.

Love-in-a-Mist, Daisy, Poppy, Honeysuckle, Strawberry, Forget-me-Not, Flax, Jessamine, Blackberry, Virginia Creeper, Hawthorn, Daffodil, Cowslip, Cherry, Buttercup, Mountain Ash, Ragged Robin, Potentilla, Apple Blossom, Strawberry and Blossom, Christmas Rose, &c. &c., also many conventional designs.

CURTAIN BORDERS.

Sunflower, Pomegranate, Passion Flower, Taxonia, Poppy, Lilies, Magnolia, Orange, Hops, Marguerites, Love-in-a-Mist, Wild Rose, Arbutus, Chrysanthemum, Iris, Cowslip, Primrose, Apple, &c. &c.

The same Designs can be had in Horizontal Borders for Chimney Valances, wide Table Borders, and can be adapted for any purpose.

N.B.—The Royal School of Art-Needlework has no Branch School nor any Agency in London.

[Pg 104]

Royal School of Art-Needlework.

EXHIBITION ROAD, SOUTH KENSINGTON.

September, 1878.

The Committee of Management of the Royal School of Art-Needlework has now organized Classes for Teaching Ornamental Needlework at their premises in the Exhibition Road, South Kensington.

These Classes are especially established for the instruction of Ladies and Children, and include every kind of stitch in Crewel, Silk, and Gold.

Ladies who wish to take lessons, or send their Children, are requested to send their names to the Secretary, who will inform them when to attend.

Each Course will consist of Six Lessons.

CREWELS.

Third Class — Six Lessons.

	£ s. d.
One Person	1 4 0
Two of same Family	1 16 0
Three ditto	2 8 0

SILK AND APPLIQUÉ.

Second Class — Six Lessons.

One Person	1 10 0
Two of same Family	2 5 0
Three ditto	3 0 0

ECCLESIASTICAL EMBROIDERY.

First Class — Six Lessons.

One Person	2 0 0
Two of same Family	3 0 0

Three ditto	4 0 0	

Single Lessons.

One single Lesson (for 1 hour) on Lesson day	0 7 0
Ditto ditto Special day	0 8 6
Ditto on Ecclesiastical Work (at any time)	0 10 6

Private Lessons at Home, 10s. 6d. the hour and expenses.

Special terms for Classes of Twelve and upwards.

[Pg 105]

FINISHED WORK.

Curtain Borders, on Serge or Diagonal Cloth, from £2 10s. to £10 10s., about 3½ yards long.

Dress Borders, on ditto, from 7s. to 18s. per yard.

" " on House Flannel, from 3s. 6d. to 10s. 6d. per yard.

Curtain Borders, on Linen, from £1 10s. to £6 6s. each.

Table Borders, on Linen, from £1 1s. to £2 10s.

Chair Backs, on Linen, from 14s. 6d. to £2 10s.

Sofa Backs, on Linen and Silk, from £2 2s. to £10.

Table Covers, on Linen, from £1 3s. to £5.

" " Serge, from £1 1s. to £7.

" " Diagonal, from 30s. to £26.

Small Chair Seats, on Diagonal, from 13s. to £2 12s.

Large " " Serge, from 13s. to £3 3s.

Cushions, made up, from £2 2s. to £5 7s.

Children's Dresses, from £1 1s. to £3 3s.

" Aprons, from 12s. 6d. to £1 1s.

Children's French Blouses, 18s. 6d. to £2 3s.

Ladies' Lawn Tennis Aprons, from £1 5s. to £3 10s.

Linen D'Oyleys, from £2 7s. to £8 8s. per dozen.

Tea Cosies, on Diagonal, from 16s. 6d.

Kettledrum D'Oyleys, each 5s. 6d. to 16s. 6d.

Sachets, with Mat to correspond, on Linen, from £1 6s.

Folding Screens, from £13 to £100.

Curtains, on Serge or Linen, from £10 to £60 per pair.

Mantel Valances, from £2 2s. to £10 10s.

Banner Screens, from £1 10s.

Counterpanes, from £6 to £80.

Table Screens, from £4 4s.

Ladies' Algerian Hoods, from £3 to £10.

Fans, Mounted, from £2 7s. to £20.

Carriage Rugs, from £2 to £10.

Blotter and Envelope Box, from £8 8s.

" on Linen, from £1 5s.

Envelope Box, on Linen, from £3.

Photograph Frames, from £1 10s.

[Pg 106] Bellows, from £1 17s.

Opera Cloaks, from £3 3s.

Nightingale Dressing Jacket, from £2.

Bath Slippers, from 6s. 6d. per pair.

Washstand Backs, from £1 5s.

Blanket Mats, for Bath, 15s. 6d.

Berceaunette Covers, from £1 10s.

Sunshade Covers, from £3 3s.

Piano Panels, from £1 3s.

Babies' Head Flannels, from £1 3s.

" Cloaks, from £4 4s.

Handkerchief Sachets, from £3 3*s*.

Knitting Pockets, from £1 1*s*.

> *P. O. Orders Payable to L. Higgin, Exhibition Road.*
> *Not more than 18 Stamps received.*

AGENTS IN THE COUNTRY.

Liverpool: Messrs. Rumney & Love, Bold Street.
Manchester: Messrs. E. Goodall & Co., King Street.
Leeds: Messrs. Marsh, Jones, & Cribbs.
Norwich: Messrs. Robertson & Sons, Queen Street.
Glasgow: Messrs. Alexander & Howell, 108, St. Vincent Street.
Birmingham: Messrs. Manton, Sons, & Gilbert.

And for

America: Messrs. Torrey, Bright, & Capen, Boston.

BRANCH SCHOOL FOR SCOTLAND:

116, St. Vincent Street, Glasgow.

All information to be obtained at the Show Rooms,
108, St. Vincent Street.

As advertisements have from time to time appeared in various newspapers offering for sale designs of the Royal School of Art-Needlework, the Public is requested to note that no designs either on pricked paper, or in any other form than on commenced work, are, or ever have been, sold by the School, or supplied to any agent. Further, that no tracing powder is used in preparing the patterns, or sold for that purpose. All designs, therefore, offered as those of the Royal School are either entirely spurious, or are pirated from theirs.

CHISWICK PRESS:—C. WHITTINGHAM, TOOKS COURT,
CHANCERY LANE.

Transcriber's Note

Minor typographic errors in punctuation have been corrected without note.

Hyphenation has been made consistent in the main body of the text without note. Please note that the author uses the term 'high light' rather than the more usual 'highlight'.

The following amendments have been made:

Page 15—grounds amended to ground—"As ground for embroidery it has an excellent effect."

Page 53—the page reference to Burden stitch has been amended from 49 to 50.

The first 10 captioned illustrations (starting with "No. 1.—Stem Stitch") have been made consistent with the later illustrations, by the removal of the word Illustration and a comma at the beginning of each of those captions.

The two illustrations on page 81 (Plates 11 and 12) were printed in reverse order in the original. They have been swapped over so they are now in the correct numeric order in this e-text.

Omitted page numbers refer to blank pages in the original.

The final section of the book (starting on page 97) included some headings in a gothic-style font. You may wish to adjust the fantasy font setting in your browser to reproduce this styling.

www.ingramcontent.com/pod-product-compliance
Lightning Source LLC
Chambersburg PA
CBHW031446210526
45464CB00005B/2344